Stairway to the Mind

Alwyn Scott

Stairway to the Mind

The Controversial New Science of Consciousness

COPERNICUS
AN IMPRINT OF SPRINGER-VERLAG

QP
411
.S46
1995

Copyright © 1995 by Springer-Verlag New York

All rights reserved. No part of this publication may be reproduced, stored in a
retrieval system, or transmitted, in any form or by any means, electronic,
mechanical, photocopying, recording, or otherwise, without the prior written
permission of the publisher.

Copernicus
Springer-Verlag New York, Inc.
175 Fifth Avenue
New York, NY 10010

Cover image courtesy of Kazuyuki Hashimoto / PHOTONICA.

LIBRARY OF CONGRESS CATALOGING-IN-PUBLICATION DATA

Scott, Alwyn.
 Stairway to the mind : the controversial new science of consciousness / Alwyn Scott.
  p.   cm.
 Includes bibliographical references and index.
 ISBN 0-387-94381-1
 1. Consciousness. 2. Neurosciences. 3. Science—Philosophy.
4. Mind-brain identity theory. I. Title.
 QP411.S46   1995
 153—dc20   95-5932

Acquisitions editor: Robert C. Garber
Manufactured in the United States of America
Printed on acid-free paper

9 8 7 6 5 4 3 2 1

ISBN 0-387-94381-1

This book is dedicated

to Lynne,
who helped me keep going
when the going got tough,

and to Lela,
who looks to the future.

Preface

Science is widely discussed in our culture. This is entirely appropriate because the gathering and ordering of information about nature has been a primary activity of the Western world since—in the nineteenth-century metaphor of Henry Adams—the Dynamo replaced the Virgin as our guiding spirit. But the nature of science is not well understood by the general public. Although the popular press presents it as an almost magical source of knowledge and predictive power that is practiced by a modern priesthood and—from its laboratories and abstract theories—radiates pronouncements on the nature of disease, colored lights, genetic codes, causes of crime, the weather, life and death, mental instability, pollution, earthquakes, the fundamental structure of matter, and the first fifteen femtoseconds in the history of the universe, the average scientist is as confused about such questions as anyone else. Grade-school teachers tell their classes that a scientist makes a hypothesis, takes some data, and discards it (the hypothesis, not the data) if it doesn't fit. Members of Congress are concerned about scientific miscreants who record false measurements, steal ideas, put their names on papers that they don't understand, and harass their subordinates. Young physicists try to find jobs. Working scientists ignore most of this; they are far too busy writing the next paper and seeking financial support for their graduate students and the laboratory animals.

Individual scientists are often not well informed about the large-scale, hierarchical nature of science because they are too close to it. To become a competent scientist, after all, is hard work. In order to be successful it is necessary to specialize in a *branch* of science, so in a sense there is no such thing as a "scientist." There are only biochemists or ethnologists, atomic physicists or chemists, electrophysiologists or psychologists. This division of intellectual labor works well in branches that can be treated as isolated nests of theory and experiment, but many questions are too broad for a single branch of science. Examples of such broader questions include:

- What is the nature of life?
- What is the nature of the mind?
- What is the nature of human existence?

These are big questions, and I hasten to add that the aim of this book is not to answer any of them. Instead it is to sketch the structure of science in a way that may contribute to our understanding of the second one: What is the nature of the mind? Thus this book is simultaneously about the science of consciousness and about the nature of science itself as it is conducted in our era.

The scientific community to which this book is directed is defined in a broad sense to include applied mathematicians, physicists, chemists, biochemists, cytologists, electrophysiologists, neuroscientists, psychologists, physicians, ethnologists, and philosophers of the mind, so questions of appropriate notation become acute. The mathematical formulas that physical scientists find helpful to describe some aspects of the story are organized in appendices so those who are familiar with this language can check the details, but the main text is written in a style that—it is hoped—will appeal to all. Nonetheless the general reader should be forewarned on one point: several of the key ideas presented in the book involve counting with rather large numbers and these ideas need to be understood. Such discussions should not be skipped over or the main thrust of the argument may be missed.

A book about the philosophy of science and the study of consciousness is likely to encounter certain questions. Why is such a book needed? What claims have I to a clearer perspective than those who have discussed this subject before? In defense against the charge of arrogance, it should be emphasized that nothing presented here is new. Each piece of the story is part of the scientific tapestry, which has been carefully woven—and often rewoven—over the past few centuries. My contribution is to step back for a look at the broad picture and suggest some opportunities for additional embroidery.

What tools do I bring to the task? Mainly those gathered during forty years of an untidy intellectual history. With an undergraduate degree in physics I spent several years working as an engineer, designing electron beam devices (traveling wave tubes and backward wave oscillators), transistor circuits, and novel microwave antennas. Motivated by an interest I developed during this period in the design of brains, I returned to Academe with the aim of understanding systems with many parts. Nowadays these would be called "complex systems." My doctoral research involved theoretical and experimental studies of a newly invented nonlinear device (the semiconductor tunnel diode) with a large area that allowed wave effects to enter the picture. In the spirit of the times my analysis was linear, but the experiments were not. One afternoon Professor Jun-ichi Nishizawa from Tohoku University (he is now the president of Tohoku University) stopped by the laboratory, saw what I was doing, and said, "In Japan people are working on the same problem, but they are using nonlinear wave theory." It was one of those defining moments so unpredictable and so important for the course of future events.

Nonlinear wave theory. The connection with the dynamics of nerve impulse propagation soon became clear, and I was launched on a program of study—lasting now for well over three decades—that has involved the development and applications of nonlinear theory to solid-state devices (semiconductors, lasers, and superconductors), nerve propagation, soliton dynamics, brain studies, quantum effects, and biochemistry. Over the years there has been a general tendency to reduce the scale of interest: For understanding the brain one must appreciate the neuron; for the neuron one must know the axon; for the axon, the active nerve membrane; for the membrane, the intrinsic membrane proteins; for proteins, the nature of amino acids; for amino acids, chemistry; for chemistry, atomic physics; and for atomic physics, the quantum theory. At each level new aspects of reality emerged, and over the years the hierarchical structure of knowledge—which is central to the message of this book—became an essential part of my theoretical perspective.

But there is another strand to the story. In my first year of college all were required to take a course in the philosophy of Western civilization that began with a reading of Ruth Benedict's classic *Patterns of culture*. This book greatly influenced my thinking, not only by its graceful presentation of a novel idea, but also from Franz Boas's introduction, which includes the following poem by Gœthe:

> *Wer will was Lebendig's erkennen und beschreiben,*
> *Sucht erst den Geist heraus zu treiben,*
> *Dann hat er die Teile in seiner Hand,*
> *Fehlt leider nur das geistige Band.*

In a pallid translation, this becomes

> *One who would know a living thing*
> *Tries first to drive its spirit out,*
> *Then with the pieces in his hand,*
> *He lacks its unifying bond.*

Quoted by Boas as satire on the position of the cultural anthropologist, this poem suggested to me—over the years—a paradigm for answering the three broad questions listed above.

Tucson and Lyngby

Acknowledgments

The ideas expressed in this book have developed slowly from interactions with many people. First of all I must thank Eduardo Caianiello, whose vision led to the founding of the Laboratorio di Cibernetica in Arco Felice near Naples with the aim of bringing together scientists from a wide variety of professional disciplines for collaborative study of the brain. Alberto Monroy provided a similarly stimulating environment at the venerable Stazione Zoologica, also in Naples. Throughout the 1970s, the Laboratorio and the Stazione gave me the opportunity to participate in discussions of the nature of the brain and the mind with many friends, including especially Antonio Barone, Renato Capocelli, Stefano Levialdi, Aldo de Luca, Luigi-Maria Ricciardi, Settimo Termini, and Uja Vota-Pinardi. Visitors from around the world enlivened these discussions, and I am particularly grateful to have had the opportunity to meet and talk with Shu-hei Aida, Michael Arbib, Jack Cowan, Stephen Grossberg, Erich Harth, and Charles Legéndy in many a *trattoria* on many a sunny afternoon.

My years at the University of Wisconsin as a participant in the Neurosciences Program were similarly blessed by opportunities to interact with Deric Bownds, Gail Carpenter, Charles Conley, Alexander Davydov, Chris Eilbeck, Dan Geisler, David Green, Chris Jones, Steve Luzader, Mike McClintock, Dave McLauglin, Bob Parmentier, and Tony Stretton, and by discussions and correspondence with Kenneth Cole, Paul Fife, Richard FitzHugh, Donald Hebb, Mark Kac, Joe Keller, Jerry Lettvin, Margaret Mead, Robert Miura, and John Rinzel. A memorable summer spent at Tohoku University as a guest of Risaburo Sato and the Japan Society for the Promotion of Science provided the opportunity to meet Shun-ichi Amari and to visit the laboratory of Jin-ichi Nagumo, where flocks of birds and schools of fish were studied along with artificial neurons, electrochemical nerve nets, and a game called *go*. Acknowledgements from the Wisconsin years would be incomplete without thanking Ivana Spalatin for introducing me to many of the philosophical ideas underlying the nature of the mind.

Over the past fifteen years at the Los Alamos Center for Nonlinear Studies, the Applied Mathematics Program at the University of Arizona, and the Institute for Mathematical Modelling at the Technical University of Denmark, I have had interesting and useful conversations with Ross Adey, Nils Baas, Mariette Barthes, Lisa Bernstein, Irving Bigio, Alan Bishop, Giorgio Careri, Peter Christiansen, Doyne Farmer, Hans Frauenfelder, Stuart Hameroff, Nitant Kenkre, John Kessler, Aaron King, Jim Krumhansl, Scott Layne, Alan Luther, Victor Mark, Neil Mendelson, Erik Mosekilde, Lis Nielsen, Roger Penrose, Michel Peyrard, Stephanos Pnevmatikos, Steen Rasmussen, Doug Stuart, Terry Triffett, Jack Tuszynski, Stan Ulam, Art Winfree, and Alex Zolotaryuk. Each of these individuals has helped me grope my way toward the perspective of this book. Those who have made specific suggestions for improving the manuscript include Deric Bownds, Karin and Peter Christiansen, Jonathan Coles, Chris Eilbeck, Stuart Hameroff, Per-Anders Hansson, Ezio Insinna, Aaron King, Lynne MacNeil, and Tish Newell, and I am particularly grateful for all of this assistance. The figures have been imaginatively and enthusiastically prepared from an odd collection of scattered items by Cindi Laukes, Dyan Louria, and Steve Uurtamo.

Thanks are also due to the National Science Foundation for three decades of support on every aspect of this work, and also to the Wisconsin Alumni Research Foundation, the Army Research Office, the National Institutes of Health, the Department of Energy, the Arizona Research Laboratory, and the Technical University of Denmark for helping out at various times in a variety of ways.

Finally it is a pleasure to thank Rob Garber, Senior Science Editor at Springer-Verlag in New York, for his encouragement, many suggestions, and support, and also for making available the assistance of Mark Uehling and Jessica Downey with the final preparation of the manuscript. Special thanks go to Jonathan Harrington for an outstanding job of copy editing. Without their help, this would have been a very different book.

Contents

List of Figures

List of Tables

Introduction

*Reality is more fluid and elusive than reason,
and has, as it were, more dimensions
than are known even to the latest geometry.*

George Santayana

I magine a wooden ladder. It leans against a tree in an orchard,
wind-battered and gray, its wood etched in dark and narrow
grooves. Though the elements have stripped it of its shine, the
wood is oddly soft. Looking at the ladder, one knows that one is *aware* of
it, as one is aware of the orchard and the fruit ripening in the trees and the
geese flying overhead.

Taking delight in our own everyday consciousness, in our simple realization that we *are* indeed experiencing a moment of joy or a hot bath, is one of
the principal pleasures of human existence. But what, we ask, is the nature
of this subjective experience that we call consciousness?

From the first days of civilization, philosophers, artists, and physicians
have tried to plumb the mysteries of consciousness. Usually they have examined some fascinating but fleeting facet of their inner lives. Plato and René
Descartes, William James and Henry James and James Joyce, not to mention other luminaries, have attempted to explore the nature of consciousness
without gathering much evidence that could be confirmed by anyone else.
This is not to say their work was unworthy: many of their ideas continue to
shape the modern quest. But intuitive investigations seldom yield observations that can be proven dispassionately.

In the last few decades, however, science has made some progress in gathering objective information about a phenomenon that is thought by many
to be ineffable. Once off limits to serious researchers, consciousness is again
becoming an acceptable subject of scientific inquiry. It has benefited from
medical technology to analyze the brain—positron emission tomography
is but one example—and still more insights come from physics, chemistry,
biology, neuroscience, psychology, and even sociology and philosophy. The
evidence, bit by bit, is derived from experiments and theories testing ele-

1

ments of the experience of consciousness, from the perception of a wooden ladder to the working of parts of the brain that are too small to be examined without the aid of an electron microscope.

Yet here, as with the efforts of ancient sages, no comprehensive understanding of consciousness has arisen from the scientific Balkanization of the subject. The research has not yet been synthesized into one overarching understanding. The experience of consciousness is richer; its explanation, by necessity, must be more complex. Consciousness cannot, alas, be reduced to the response to an inkblot or the activity of a set of neurons. Yet too often have scientists decapitated the reality of consciousness with Ockham's razor, the metaphorical blade by which everything is cut to its simplest essence.

This scientific tendency to reduce nature to its elements may have reached bottom at a leading university, which cannot be named, where a well-known physicist labored for years, and finally announced he had developed a theory of everything—an elegant conception long sought by his rivals. His theory explained, to the last detail, the motion of every particle in the universe. Even better, the theory could be reduced to an equation: E, for everything in the world, equals zero. Sadly, however, the physicist informed his peers that he needed 243 volumes (and an appendix) to describe E.

In this more modest work, consciousness will not be reduced to an equation. If we handle Ockham's razor more judiciously, leaving the head of consciousness attached to the complicated body of everyday experience, we must first acknowledge those who still believe consciousness can be reduced to some as-yet-undiscovered cluster of cells in the brain, or to some heretofore overlooked application of the laws of physics. According to this school, the future is determined by the present, and the present by the past. Such thoughts are known as the reductive materialist view.

There is also the related functionalist position, which holds that the essential aspects of mental dynamics will eventually be expressed as a formula and represented on a system constructed from integrated computer circuits. Since our consciousness is based on the world of cells and atoms, it might one day be represented in silicon chips as well. But both the reductive materialists and the functionalists seem to depreciate our multidimensional and intensely subjective awareness of a ladder in an orchard.

The main philosophical alternative is also unsatisfactory. The dualist camp, as it happens, believes that consciousness somehow transcends the physical levels of the brain. It is undeniable that our psychic experiences—dreams and memories, feelings and premonitions—are integral to consciousness. Nor can it be disputed that such experiences are mysterious, or that science offers little explanation. But to divorce such phenomena from the physical realm is to regress into the hazy prescientific era of myths and fables and legends.

This book proposes a different approach. Without resorting to myth, without confining myself to the quantifiable physical world or disavowing it, it is possible to illuminate what can be known—scientifically—about our awareness of the world. Moved by the image of the ladder in the orchard, I propose a related image, a metaphorical stairway, that is equally practical. This stairway is a hierarchy of mental organization in which most of us, as we go about our lives, unwittingly stand at the two top steps or levels: consciously aware of the realities of our culture. But the lower steps are every bit as important to our position as the uppermost level. (If you don't agree, try taking them away!) Thus, I suggest, consciousness is an *emergent* phenomenon, one born of many discrete events fusing together as a single experience.

Most adults experience their culture's collective understanding of what it is to be conscious. Through the mind, they may also be conscious of an identity, a profession, a gender, a social class, a place, a body. Thanks to the peculiarities of their culture, they may be conscious of love, anxiety, or dread. Below the two top steps of the stairway is the basic biological seat of consciousness: the brain. And below that level sit the subcomponents of the brain, assemblies of tangled cells (called neurons) bearing the weight of subtleties to which the conscious mind, on the steps above, is not often privy. One might imagine the upper levels of this hierarchy thus:

Culture
Consciousness
Brain
Assemblies of neurons
Neurons.

Smaller than the neurons, themselves made up of even smaller structures, other biological building blocks occupy the lower steps of the stairway. Hence the lower levels of the hierarchy might include:

Nerve impulses
Biochemical structures
Molecules
Atoms.

This hierarchical scheme is by no means original; it is based upon equal parts of scientific observation and common sense. What's more, it is not definitive or all-inclusive. It might be constructed of perhaps a dozen additional steps, depending upon the perspective of an observer or the specificity of one's knowledge. The point is that to understand one level of the hierarchy, the levels below must be probed. To understand consciousness, one must explore the brain; to understand the brain, one must investigate the neuron; to understand the neuron, one must study the dynamics of proteins; to understand proteins, one must become familiar with chemistry and atomic physics.

Put another way, one can go from 34th Street to the top of the Empire State Building like a tourist, taking the express elevator and skipping all the intervening floors in order to concentrate his attention on the view of Manhattan. Although this approach saves both time and effort, the tourist learns nothing about the building itself. On the other hand, one might participate in the Annual ESB Stair-Climb, taking time to look at each floor along the way and develop an appreciation for the complex nature of a modern office building. In the case of consciousness, naturally, any passage from one level to the next is more complicated than climbing a flight of stairs in the Empire State Building; the components on one level of the hierarchy coalesce into something more organized, something qualitatively different, at the next higher level.

To understand the relation between one level and the next, let us ignore the tendrils and filaments of neurons, to which we shall return soon enough. Instead let our thoughts return to the ladder in the orchard on a warm autumn day. From the northwest, a cloud bank carries cold air over the humid landscape below, pushing the pale blue sky to the southeast, where a prudent flock of geese disappears into the distance. Soon the sky is covered with a carpet of greenish gray—like a week-old bruise—punctuated by occasional flashes of lightning. Close to the ground nothing stirs. The air is still; small animals are huddled safely in their burrows. Curls of cloud begin to pop out of the ugly carpet and reach toward the ground, feeding on the energy that is stored in the warm, wet air below. Suddenly a single curl drops to the ground, gulps an amount of energy that staggers human imagination, and shatters the stillness, sucking trees and bushes into its hungry maw. Moments later rain fills a nearby river above its banks, cutting a new channel through the hills. A young woman leans against the rungs of the ladder, watching all.

Clearly the tornado emerges not as a preordained event—caused by the heat, say, or the wind—but out of many ingredients mixing together. In the world of meteorology this is always the case. One gust of wind plus one warm day does not necessarily add up to a cyclone.

In the terminology of applied mathematics, a tornado is a *nonlinear* event, one in which the whole is more than the sum of its parts. In traditional *linear* mathematics, one may remember from the classroom, it is a simple matter to predict how far a certain weight will stretch a spring, or where an artillery shell will land. The spring or trajectory, in such cases, is *exactly* the sum of its parts. But in the realm of nonlinear mathematics, one needs a supercomputer to predict when, or even if, a tornado will arise from a thunderstorm. And even then, as we know, the prediction may be wrong.

For that reason, linear phenomena have long been favored by scientists. Ricocheting billiard balls and swinging pendulums are often described because we understand them so well. But nonlinear phenomena are by far the more natural and commonplace in biology. Three more examples of nonlinear phenomena must suffice for the moment. One goose plus another plus another plus a dozen more equals...not fifteen geese but something altogether more organized, a *flock*, with its own peculiar behavior, moving as a single organism, V-shaped in its flight. The channel cut by the deluge on that autumn afternoon is shaped and broadened by later rains to become a new branch of the river. Many years later, the young woman's experience is recreated in a poem. The tornado and the flock, the channel and the poem are all nonlinear events, and so they cannot be easily calculated or predicted from a few simple starting conditions.

It should be no surprise that consciousness, a biological reality even to children, is yet another instance of such nonlinear coherence. Consciousness *emerges*—like a raging tornado or a gentle flock of birds. Just as life emerges from several of the lower levels of the hierarchy, consciousness emerges from several of the upper levels. To appreciate fully the nature of consciousness, one must consider several steps of the stairway together.

Perhaps the most intriguing aspect of the stairway to the mind is its ability to lead from one area of scientific research to another. The scientific exploration of consciousness is at such an early stage that researchers studying different levels of the hierarchy have not yet committed themselves irrevocably to a single step. In meetings around the world, neurologists have presented experiments to explain volition, impressing philosophers. Anesthesiologists, by the same token, have succeeded in reaching patients in surgery with the word games of psychologists. Moving down the stairway, biologists have been borrowing ideas from quantum mechanics and physical chemistry.

All these scientists, from biochemists to ethnologists, have shown that the hierarchy of consciousness has more than a few levels. No doubt others remain to be discovered. But enough have been identified and tested so that it is possible to begin with an everyday conception of consciousness—a young man staring at a young woman and happily *aware* he is staring at her—and move slowly toward an appreciation of the cellular and physical events that make our awareness vivid.

Science, to be sure, does not have all the answers. But we have begun to make our way up the stairway, to make progress that may be interesting not only to scientists in the field but also to less technically minded readers. This book is intended for both audiences in the spirit of William James, a pioneer in the both the study and the appreciation of human consciousness. "As the brain changes are continuous," James wrote, "so do all these consciousnesses melt into each other like dissolving views. Properly they are but one protracted consciousness, one unbroken stream."

Reference

G Santayana. *The sense of beauty.* Dover Publications, New York, 1955 (first published in 1896).

Quantum Physics, Chemistry, and Consciousness

*The ability to reduce everything to simple fundamental laws
does not imply the ability to start from those laws
and reconstruct the universe.*

Philip W. Anderson

A glassblower who wishes to create an exotic vase that one day will be filled with a beautiful arrangement of flowers does not begin by venturing into his garden to cut roses. Similarly, physical scientists considering consciousness should not start out by discussing the brain of William Shakespeare. First, like a glassblower heating a supply of pure white sand, they must consider more elemental ingredients. These ingredients, protons, neutrons, and electrons, are the materials on the workbench of a physicist. Physicists have proudly fashioned a vase from such sand and passed their glowing creation along to chemists, who, like florists, are supposed to gaze upon it with the utmost reverence. But the chemists, again like florists, have developed their own uses for the glass, filling it with all manner of things, and their rules may not ultimately derive from the respectable traditions of glassblowing.

Perhaps the greatest glassblower of all, a hero not only to physicists and chemists, was Isaac Newton (1642–1727), the Englishman who established the elemental laws of the world we can easily observe. Newton remains the preeminent example of a philosopher-scientist who showed how the momentum of one billiard ball struck by another could be precisely calculated and explained the mystery of planetary motion on the basis of a few simple laws. His ideas, now three centuries old, ushered in the classical age of physics, a time of great scientific certainty. In the modern world, as we all know, there are phenomena that do not conform to Newton's laws and cannot be explained as readily as a falling apple.

These non-Newtonian events are of particular concern to the student of consciousness because our brains, beyond all doubt, obey the same physical laws as apples and molecules of oxygen. But a central question is this: What other laws, among all those formulated over the past three hundred years, govern the dynamics of a brain? And what of the mind, that mysterious locus of conscious awareness and feelings? The modern scientist wonders whether our minds operate entirely in accordance with the laws of motion as prescribed by Newton, or also obey some other yet undiscovered set of principles.

It is entirely clear, of course, that our brains process and transmit signals in accordance with rules familiar to physical chemists and electrical engineers. And since nearly a century of investigation into events at the molecular level has produced considerable insight, that knowledge—in theory—might be applicable to problems like consciousness. But, as we shall see, some scientists believe that the most basic computational power of the brain derives from certain structures, *quantum* structures, that do not follow the laws of classical physics. Before we can examine that particular view, we must review some of the concepts applying to that realm—particuarly those from the post-Newtonian era, laws relating to entities much smaller than neurons.

QUANTUM PHYSICS

Toward the end of the nineteenth century, leading European physicists were busily working out the rules for rarer, more abstract events than falling apples. One such phenomenon was the radiation from a *blackbody*, an object that reflects no energy. A simple example of a blackbody is provided by our glassblower's oven. Let us think about it in the early morning, when it is still cold, and imagine a small hole that is drilled through one side. Any light that enters this hole is absorbed during multiple reflections from the walls so almost none escapes and it looks black. This little hole in an oven is a very close approximation to a physicist's ideal blackbody.

When cold, the blackbody is completely dark, but as the oven warms, it begins to glow. At first this radiation is only felt by the cheek and not seen by the eyes because it lies in the infrared range of frequencies. At progressively higher temperatures, the radiation becomes a dull red, then orange, and finally white as the sun. Nineteenth-century physicists knew that the temperature of the glassblower's oven can be rather accurately measured by comparing its color with a color chart similar to those in a paint shop. By the end of the nineteenth century, the color (or *spectrum*) of blackbody radiation had been carefully measured as a function of temperature, but it could not be calculated as a function of temperature from the laws of classical physics.

It fell to Max Planck (1858–1947), a German physicist, to show how the color of blackbody radiation could be calculated. To perform this calculation, he assumed that the inside of the oven could be represented by a large number of tiny atomic and molecular oscillators: little masses connected to springs vibrating at many different frequencies. As the temperature increased, his theory showed how the average energy—and thus the color—of this collection of tiny oscillators would change.

The experimental measurement that Planck strove to predict from first principles (the color of a glassblower's oven as a function of its temperature) was well known at the time, and he continued to get an incorrect result.

Finally in 1901 he felt forced to make an assumption that seemed wrong. Under this premise, the vibrational energy of each of the myriad oscillators within the oven could not take on *any* value. Only certain energy levels were allowed, but that assumption led to a prediction in perfect agreement with measurements.

It was as if the energy of each oscillator existed in little packets that came to be called *quanta*, the Latin term for "How many?" Planck obtained the difference between two energy levels of each tiny oscillator by multiplying the frequency of an oscillator (or the number of times that it alternates each second) by a very small constant number. This small number, which is equal to 6.626 times 10^{-34} joule-seconds, will appear again in relation to our discussions of consciousness. It came to be called *Planck's constant* and won him the Nobel Prize in 1918, but Planck was not happy with his result. Why should the energy of an oscillator take only certain values? It seemed a crazy idea, which he had accepted only because he could see no other way to correctly derive the color of a glassblower's oven from its temperature.

Eight years after receiving the Nobel Prize, Planck was sent six papers from another physicist, the first of which arrived at *Annalen der Physik* on January 27, 1926 (Schrödinger, 1926). Upon receiving this manuscript, Planck replied that he had read it "like an eager child hearing the solution of a riddle that had plagued me for a long time."

Unlike Planck's own work on blackbody radiation, this paper considered the simplest chemical element in the universe: the hydrogen atom, which consists of a single positively charged proton circled by a single negatively charged electron. The author of the manuscript was Erwin Schrödinger (1887–1961), an Austrian physicist, philosopher, and sometime poet, who had happened to pass the previous Christmas vacation with a mysterious lady friend at a villa near Davos in Switzerland. When later asked if he had enjoyed the skiing, he replied that he had been distracted by a "few calculations." These calculations predicted the frequencies at which a single hydrogen atom emits light. (Mathematically minded readers can find the equations in Appendix B.)

Supposing the relatively lightweight electron to be rotating about the proton, much as our earth circles the sun, one might expect the hydrogen atom to gradually lose energy. A falling object eventually comes slowly or abruptly to rest; a moving vehicle eventually ceases to move. If one could chart the level of energy within the hydrogen atom, one might expect a similar emission of a broad band of radiation as the electron wound to a stop. But what the experimentalists find is altogether different. They observe a line spectrum—a series of sharply defined peaks—rather than a continuous band at all wavelengths. In this respect, as with the tiny oscillators in Planck's oven, the actual results lie at discrete levels, not along a continuous curve on a graph.

To appreciate the importance of Schrödinger's results for consciousness, let us assume that we are attempting what might be a comparatively simple task: to determine the precise path (or trajectory) of an electron circling a proton in the brain of William Shakespeare. There are two ways this might be accomplished, and since we are talking about the same electron in the brain of the same person, one might expect that their answers would concur. From the perspective of *classical* physics, the physics of Isaac Newton, the problem of calculating the electron's trajectory is *nonlinear* in the sense that is defined in Appendix A. In a linear problem, the sum of two causes yields an effect that is the sum of the two individual effects. In a nonlinear problem, this is not the case. Although classical physics generates both linear and nonlinear problems, the more interesting ones, like a calculation of an electronic trajectory in William Shakespeare's brain, are nonlinear.

Now from the perspective of *quantum* physics, as it was conceived by Schrödinger, problems are always linear, but the prediction of an electron's trajectory becomes tenuous. The quantum physicist begins by solving Schrödinger's equation for certain special functions, called *eigenfunctions*, each of which has a corresponding value of energy called its *eigenvalue*. Since Schrödinger's equation is linear, any number of eigenfunctions added together—even an infinite number of them—will also be a solution of the equation. Such a sum is called a *wave packet*, and it replaces the classical trajectory as the basic concept of quantum mechanics.

Each component of a wave packet (each eigenfunction) has a particular frequency of oscillation, which is equal to its energy divided by Planck's constant. Thus the entire packet is rather like the sound that emerges from a piano, where each string is vibrating at its own frequency. In this analogy, the vibration of a piano string corresponds to the oscillation of an eigenfunction of Schrödinger's equation. In order to understand the nature of a wave packet, one must become familiar with its components: the eigenfunctions.

At this point the reader may be asking: What *is* an eigenfunction? Briefly, it is an electron cloud that forms a pattern in space indicating the probability of finding the electron at various points in the vicinity of the protonic nucleus. Let us look at some of these eigenfunctions for the hydrogen atom. The eigenfunction of lowest energy—by convention called the "1s" state—is shown in Figure 1. This state is spherically symmetric so the figure is meant to look like an orange with a quarter cut out and viewed in profile.

The energy of the 1s eigenfunction is about 13.6 electron-volts below that of a free electron, and this difference is called the *Rydberg energy*.[1] (One electron-volt, the reader may recall, is the energy needed to raise an electron through a potential difference of one volt. It is equal to 1.6021 times 10^{-19}

---

[1]In fact, the experimentally measured value for the Rydberg energy is

$$R_{Exp} = 13.6056981 \pm .000004$$

electron-volts, while the theoretical value is

$$R_{Th} = 13.605698 \pm .000022 .$$

This agreement is a remarkable achievement of Schrödinger's theory. (See Appendix B for more details.)

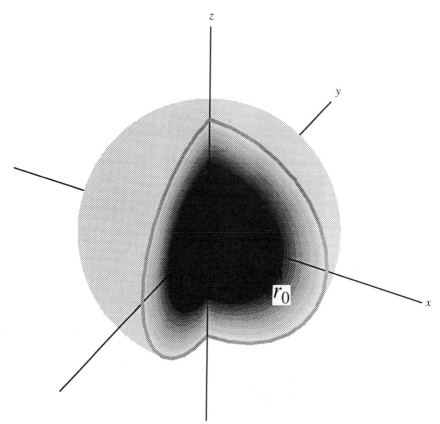

Figure 1. *The electron density cloud for the lowest-energy electronic eigenfunction—the 1s state—of the hydrogen atom. The electron remains within the spherical region for 93% of the time. Higher electron density at the center is indicated by darker shading.*

joules.) In the planetary analog of the earth orbiting around the sun, 13.6 electron-volts corresponds to the escape energy of the electron from the field of the positively charged nucleus. This eigenfunction oscillates at a frequency that is equal to its energy divided by Planck's constant.

When he first discovered these eigenfunctions, Schrödinger wondered what they meant, what they were trying to say. The German physicist Max Born (1882–1970) suggested—and his colleagues generally agreed—that the square of an eigenfunction's magnitude at some point in space gives the probability per unit volume (or probability density) of experimentally finding the electron at that particular point.

In Figure 1, the shading attempts to indicate this probability density (it is as if the orange were sweeter near the center). The radius r_0 is about 0.525 times 10^{-8} centimeters, which is a convenient unit of size for atomic structures. The 1s state, shown in Figure 1, is the only eigenfunction at an energy level of 13.6 electron-volts below that of a free electron.

Other energy eigenvalues of the hydrogen atom are found at energies of $13.6/2^2, 13.6/3^2, 13.6/4^2, \ldots, 13.6/n^2$ electron-volts, where n is an arbitrarily large integer. The differences between these energy eigenvalues correspond exactly to the spectral lines found by experimental physicists.

It is important to notice that the quantum cloud in Figure 1 and the classical trajectory of the electron cannot easily be reconciled. The two methods to calculate the location of an electron in the brain of William Shakespeare, in other words, do not agree with one another. Both theories produce predictions about the position of the electron, but these pictures are fundamentally different. The classical theory says exactly where the electron is at each instant of time (that's what a trajectory *is*), but there is no experimental evidence to indicate whether or not this calculation is correct. The quantum theory, on the other hand, does not say where the electron is. Instead it gives the probabilities of finding the electron at different points in space. While somewhat vague, these quantum predictions have the advantage of being in agreement with all experimental measurements.

At larger values of the integer n, one finds additional eigenfunctions (or electron clouds) with more interesting shapes. For $n = 2$ there is a $2s$ eigenfunction with the spherical shape indicated in Figure 1 and additional eigenfunctions called "$2p$." These are not spherically symmetric but have the shape indicated in Figure 2. There are three of them, one for each direction of space.

For $n = 3$, in addition to a $3s$ eigenfunction (similar to Figure 1) and three $3p$ eigenfunctions (similar to Figure 2), there are five additional $3d$ eigenfunctions of increased geometrical complexity. In general, Schrödinger showed, there are $2n - 1$ additional nonsymmetric eigenfunctions of ever increasing geometric complexity, where $n = 1, 2, 3, \ldots, \infty$.

Let us go over this again and focus our attention on the concept of a wave packet because some suggest that it provides a basis for understanding human consciousness. Since the quantum description of the hydrogen atom is linear—so the action of a sum is exactly equal to that of its parts—one can add eigenfunctions to obtain a composite solution. In general this wave packet is the sum of an infinite number of terms.

The first term is an arbitrary constant times a $1s$ eigenfunction (see Figure 1) with an escape energy of 13.6 electron-volts below that of a free electron. The second term has the same form as the first in order to include two directions for the *electronic spin*. The third term is a $2s$ eigenfunction (similar to Figure 1), and the fourth is another with opposite spin, both with escape energies of 13.6/4 electron-volts. The fifth through tenth terms are $2p$ eigenfunctions (as in Figure 2), also with energies of 13.6/4 electron-volts. The eleventh and twelfth terms are $3s$ eigenfunctions (again similar to Figure 1), the thirteenth through eighteenth terms are $3p$ eigenfunctions (as in Figure 2), and the nineteenth through twenty-eighth terms are the $3d$ eigenfunctions, all with energies of 13.6/9 electron-volts. And so on through the infinite number of terms in the wave packet, each of which is multiplied by

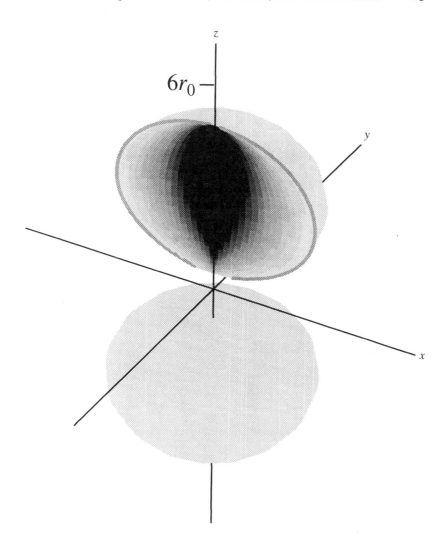

$6r_0$

Figure 2. *The electron density cloud for a 2p state of the the hydrogen atom. There are three of these states: one oriented along each of the three perpendicular axes. The electron remains within the figure indicated for about 95% of the time. As in Figure 1, the darker shading indicates regions of higher electron density.*

an arbitrary constant. These arbitrary constants are to be determined from an observer's knowledge of the initial location and speed of the electron.

Each component eigenfunction of the wave packet has a specific frequency, and components add and subtract from each other in a complicated manner, as the tones from the strings of a piano add and subtract to form the rich and satisfying chords of which we are conscious. Thus the wave

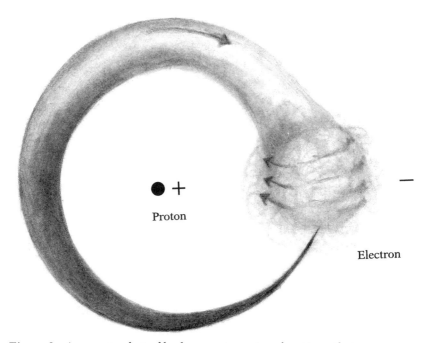

Proton

Electron

Figure 3. *A wave packet of hydrogen atom eigenfunctions that represent an electron in a classical orbit. Notice how the wave packet spreads out—or disperses—as the electron moves. (After von Hippel, 1959.)*

packet exhibits most interesting behavior. A suitably selected collection of eigenfunctions, for example, might suggest an electron rotating about the protonic center, as shown in Figure 3.

This compilation, this wave packet, is supposed to represent an observer's state of knowledge about the system being studied. But there are limits on this knowledge that were first identified by another German physicist, Werner Heisenberg (1901–1976), who in 1927 published an idea that would establish yet another distinction between the ideas of classical physics and those of quantum physics. Heisenberg was the first to understand that the light of an as-yet imaginary microscope, one sufficiently powerful to observe a proton and an electron, would necessarily interact with the atom under observation. Thus the very act of observation of an electron's position would introduce uncertainty into one's knowledge of its speed. This was a novel proposition to the Newtonian physicists, who had long assumed that everything could be measured with arbitrarily high precision.

Heisenberg's principle of uncertainty can be understood as a statement about the precision with which one can know both the speed and the position of a moving object. The uncertainty in the position of an automobile traveling along a highway, for example, might be expressed by saying that it is located somewhere between mileposts 365 and 366. In this case the uncertainty in our knowledge of the car's position would be one mile. If we know that the speed of the automobile is somewhere between 60 and 65 miles per hour, our uncertainty in knowledge of the car's speed is 5 miles per hour. Our uncertainty in the car's *momentum* would then be the uncertainty in its speed multiplied by its mass.

Specifically, Heisenberg's uncertainty principle states that the product of one's uncertainty about the position of a moving object times one's uncertainty about its momentum (mass times speed) is no less than Planck's constant. Similarly, the product of the uncertainties in the energy and time of an event must also be equal to or greater than Planck's constant.

Let's examine why this is so. If we know with reasonable accuracy the position of an electron in a hydrogen atom, then the wave packet must be composed of many different eigenfunctions. This is because, as we have seen, each eigenfunction alone says little or nothing about the location of an electron. It is only when many eigenfunctions are added together in a wave packet, as in Figure 3, that the classical trajectory of the electron begins to emerge—like the ghost of Hamlet's father—from the clouds of quantum theory. But the wave packet is limited by its very nature in the precision with which it can simultaneously fix both the position and the speed of the electron. These limitations cannot be evaded within the context of a linear quantum theory. They emerge in a natural way when our knowledge of position and momentum (or energy and time) of a moving particle is represented by a wave packet of eigenfunctions of Schrödinger's equation, because that equation is linear.

Given some initial conditions for this wave packet—that is to say, the structure of the packet at a particular time—it will evolve in a deterministic manner as required by Schrödinger's equation, and our knowledge of the position of the electron must become ever less precise because we never know exactly how fast it is going. To see this, consider, once again, our automobile passing mile post 365 on a highway at exactly 8 o'clock in the morning. At that moment, we know its position with high accuracy. But if all we know about its speed is that the velocity is in the range between 60 and 65 miles per hour, our knowledge of its position at 11 o'clock is accurate to only about 15 miles. This effect is called *dispersion*, and such spreading of the wave packet is indicated in Figure 3 by a broadening of the cloud of probability density as the electron wave packet moves. The bigger the cloud, the greater the uncertainty about the position of the electron.

Let us assume that our knowledge of an electron's position has deteriorated because of dispersion. Let's also assume that we remeasure it. At this moment, *something strange happens*. The wave packet that describes the state of our knowledge must be adjusted at the moment we look at the electron and update our knowledge of its position. This adjustment of the wave packet is referred to by physicists as *collapse*, and the phenomenon is unrelated to the dynamics of Schrödinger's equation. As Schrödinger (cited in de Broglie, 1960) put it:

There is something *psychological* about the wave packet,

which he didn't like. The great German physicist Albert Einstein (1879–1955) didn't like it either. In December of 1926, Einstein wrote to Max Born (see Ferris, 1991):

> The theory says a lot, but does not bring us any closer to the secret of the "Old One." I, at any rate, am convinced that *He* is not playing at dice.

The reader should be aware that the collapse of a wave packet is not something one sees in normal life. If it were, Isaac Newton most certainly would have discussed it. Consider, for example, a baseball that is traveling toward home plate at a speed of 90 miles per hour. Since its size multiplied by its momentum is about 0.1 joule-second, a factor of 10^{32} larger than Planck's constant, both its speed and its position can be simultaneously measured with accuracies up to one part in $\sqrt{10^{32}} = 10^{16}$ (ten million billion) without violating the uncertainty principle. Clearly this is not a restriction on the classical point of view. There is no way to detect the collapse of the baseball's wave packet because it is far too small to be measured; thus Newton's classical equations work perfectly well. For an electron in a hydrogen atom, on the other hand, the uncertainty principle imposes severe restrictions on simultaneous measurements of its speed and position.

Why were Einstein and Schrödinger so displeased with the standard interpretation of quantum theory? The answer is that they saw the theory as an incomplete or unfinished description of nature. On the one hand, Schrödinger's equation provides a wonderfully accurate description of the evolution in time of a quantum mechanical wave packet. Many verifiable predictions about the hydrogen atom can be obtained from this theory, so it evidently embodies much that is true. The collapse of a wave function, on the other hand, was said to take place in the instant that new data are registered in the mind of a conscious observer, and there was no theory to describe this phenomenon. Seventy years later there is still no such theory.

One alternative to this dissatisfaction with the structure of quantum theory has been to suppose that Schrödinger's equation might be an approximation to some not yet discovered *nonlinear* theory. Since Heisenberg's uncertainty principle—and the collapse of the wave function—arise from the linearity of Schrödinger's equation, a nonlinear version might evade the problem of a collapsing wave function (de Broglie, 1960).

Physicists have recently explored this possibility by attempting to detect a nonlinear pattern to the atomic energy levels in metallic ions. These measurements indicate that such a nonlinearity—if it exists at all—must be less than one part in 10^{20} of the Rydberg energy (Bollinger, et al., 1989; Weinberg, 1989). This is an exquisitely small amount of energy, one that appears in the biological realm to be negligible. For the study of consciousness-related problems in this book, therefore, we shall take the most conservative approach and continue to use quantum theory in the standard, linear version, as it was developed to explain the facts of atomic physics.

Classical physics, as we know, generally considers an *exact* calculation of the trajectory of a moving particle as the solution of a nonlinear dynamic problem. But the classical theory, as we have also seen, is only approximately true, becoming more accurate for situations in which the product of the energy scale and the time scale (or the momentum scale times the distance scale) becomes larger in units of Planck's constant. It is at this point that the glassblower—the physicist—has something to hand over to his colleague the chemist. For here the theoretical insights of physics lead to observations useful on a level of rather greater complexity: the description not of just a hydrogen atom but of atoms much more substantial.

For those atoms with more than one electron, it is not possible to obtain exact eigenfunctions and energy eigenvalues as Schrödinger did for the hydrogen atom. Computational difficulties arise because the electrostatic potential experienced by any one electron depends on the positions of all the other electrons. Quantum calculations that agree with measurements within experimental error have been carried out for the helium atom (with two electrons), but more severe assumptions must be made to obtain approximate solutions for the heavier atoms (Slater, 1960).

Nonetheless, quantum theory does manage to produce some useful results. This passing of the vase leads, in fact, to qualitative descriptions of the chemical elements that are observed in nature and can be organized in the periodic table (Figure 4) of Dmitri Mendeleev (1834–1907), the Russian chemist who first published his scheme for classifying the elements in 1869. To follow the construction of Mendeleev's table, it is necessary to recall that two electrons can be placed in hydrogen-like states: one for each orientation of the electronic spin, which is indicated in Figure 3.

The chemical properties of lithium (Li) are similar to those of sodium (Na), immediately to its right, because both have the same number of electrons in the outer shell. The same is true of beryllium (Be) and magnesium (Mg), boron (B) and aluminum (Al), and so on. The entry for carbon— C 6 s^2p^2—indicates that it has six electrons, two of which are closely bound in the 1s state shown in Figure 1, and the notation "s^2p^2" indicates that the four outer (or valence) electrons are in a mixture of 2s states (which are similar to those shown in Figure 1) and 2p states (shown in Figure 2). Similarly, the entry Si 14 s^2p^2 indicates that silicon has fourteen electrons, two of which are closely bound in the 1s state and eight of which are closely bound in a stable completed shell of 2s and 2p states. This leaves four valence electrons that are in a mixture of 3s states (as in Figure 1) and 3p states (similar to those shown in Figure 2). Since the s^2p^2 valence electrons of carbon are similar in structure to those of silicon, these two atoms could be expected to have similar chemical properties. This is in fact the case. A corresponding analysis can be made for germanium—Ge 32 s^2p^2—which has valence electrons in a mixture of 4s states (as in Figure 1) and 4p states (similar to those shown in Figure 2).

Perhaps the most striking property shared by carbon, silicon, and germanium is the ability to form a crystalline structure called diamond when the

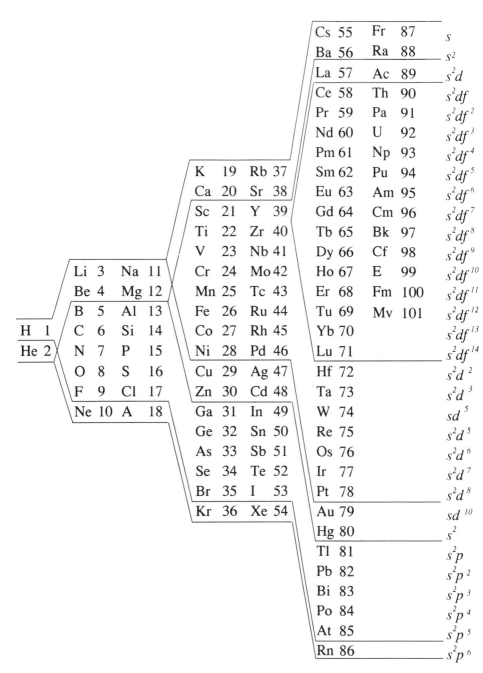

Figure 4. *Periodic table of the elements, with electronic configurations of the outer shell of electrons. The numbers indicate the total number of electrons in an atom. (After Slater, 1960.)*

atoms are carbon. Early developments in semiconductor electronics (during the 1950s) were based on the diamond structure of germanium atoms, but today the corresponding crystal that is composed of silicon atoms is almost universally used. The fact that life as we know it uses the (organic) chemistry of carbon while the computer industry uses that of silicon is both economically vital and scientifically trivial, given that diamonds play no role in biology.

The fourth column in the periodic table is somewhat more complicated because the $3d$ states become filled before the $4p$ states. Since there are five $3d$ states, a total of ten elements appear between Sc 21 s^2d and Zn 30 s^2. The fifth column is chemically similar to the fourth column, but the sixth column fills the $4f$ states before the $5d$ states, which in turn are filled before the $6p$ states. Nonetheless, columns four, five, and six include elements that are chemically related to corresponding entries in the second and third columns. Column seven starts out like column six, but the atomic nuclei tend to become unstable, leading to division (or fission) into smaller atoms. This instability leads to a limited number of atoms and provides the basis for atomic energy and the bomb.

Taken together, the 101 elements in Figure 4 are the building blocks of ordinary matter. And chemistry, one need hardly add, is the branch of science that studies how molecules can be constructed from these elements, and how those molecules interact. The American physicist John Slater (1951) has commented:

> Atoms by themselves have only a few interesting parameters. It is when they come into combination with each other that problems of real physical and chemical interest arise.

CHEMISTRY

Following Slater's suggestion, it seems reasonable to acquaint ourselves with each combination of atomic elements as thoroughly as the physicists have tried to acquaint themselves with the structures of the atoms. So let us begin by counting the number of possible molecules. First of all we ignore the noble gases (helium, neon, argon, krypton, xenon, and radon) because they have complete outer shells of electrons and therefore do not participate in chemical reactions. This leaves 95 atoms available for one molecular permutation or another. If a molecule has N atoms, we might suppose there to be 95 choices for the first atom, 95 for the second, and so on for a total of 95^N possibilities. (The number of ways that a single atom can be chosen from 95 of them is 95. The number of ways that two atoms can be chosen from 95 is 95 times 95 = 95^2 because there are 95 choices for the second atom for each choice of the first. Thus there are 95^N ways to choose N atoms.)

However, the counting is more complicated because it is necessary to consider the question of *valence*. The alkali metals from the upper row of the

table (hydrogen, lithium, sodium, potassium, rubidium, cesium, and fran-
cium) have one electron more (an s electron) than is necessary to complete
a spherically symmetric shell of the electronic eigenfunctions. These atoms
are said to have a valence of +1 because one extra electron is available to
help other atoms complete their shells. Above the noble gases are the halides
(fluorine, chlorine, bromine, iodine, and astatine), which have a valence of
-1, indicating that one electron must be added to an s^2p^5 electronic configuration to complete an s^2p^6 shell. And so on throughout the periodic
table.

As a rule, the total valence of a stable molecule will be zero (although
there are exceptions), because this lowers the quantum mechanical energy.
This valence constraint can be accounted for in a very conservative way by
counting the number of molecules that can be constructed from the first $N/2$
atoms and then choosing the second $N/2$ to have complementary valences.
Only the first $N/2$ permutations are assumed to contribute to the counting;
thus the number of molecules that can be constructed from N atoms is
greater than $95^{N/2}$.

Now $95^{N/2}$ is a rather large number, even for moderate values of N,
as a moment with the pocket calculator will demonstrate. If, for example,
$N = 112$, then $95^{N/2} = 95^{56}$, which is approximately equal to 10^{110}. (This
is a one followed by one hundred and ten zeros.)

The physicist Walter Elsasser has coined the term *immense* to describe
numbers that are larger than

$$\Im = 10^{110}.$$

His selection of this magnitude was not arbitrary. The mass of the universe
measured in units of the mass of a hydrogen atom is about 10^{80}, and the age
of the universe (20 billion years) measured in units of picoseconds (10^{-12}
seconds) is about 10^{30}. Thus the mass of the universe multiplied by its age
is roughly equal to 10^{110}. According to Elsasser (1966, 1969), any finite
number greater than 10^{110} is said to be immense.

Why is this definition of interest? Because one generally assumes that any
finite number of items can be put on a list and examined, one by one. For an
immense number of items, however, this is not possible. The largest currently
available computer—and any computer that could ever be built—would not
have enough memory capacity to store an immense list. Even if a single atom
were used to remember each item on the list, it would still be far too long.
Furthermore, if such a list were assumed to exist somewhere—say in the
mind of God—there would not be time enough to read it by any conceivable
means.

The significance of an immense number of possibilities can be appreciated
by comparing the ways that one approaches two ancient board games: tick-
tack-toe and chess. With tick-tack-toe, there are only a few dozen distinct
games, each of which can be studied in detail. From such a study, it is

straightforward to show that the the player who goes second (the ○) need never lose. The player who goes first (the ✕) can only win if ○ makes a well-defined mistake. With the game of chess, on the other hand, the number of possible games is immense. Thus there will never be a master chess book that describes and discusses all possible games, and one's strategy must include generous elements of intuition and daring. There will always be many exciting chess games that have never been played.

If we set the number of molecules that can be constructed from N atoms (which was estimated to be greater than $95^{N/2}$) equal to $\mathfrak{I}$, we find that N is less than 112. In other words, the number of molecules with more than 112 atoms is immense, so the chemist's work will never end. There will always be many interesting molecules that have never been studied.

As we move upward on the hierarchical ladder of science, we shall see that immense numbers of possible structures are ever easier to find. The possible numbers of proteins, cell types, mental states, and cultural configurations are *all* immense, so it is not possible to predict what sort of structures might emerge at these levels. It is necessary, therefore, to *look and see what is there* at each level before constructing an analytical theory.

Facing such an overwhelming number of possible molecules, the theoretical chemist attempts to apply quantum theory to study a molecular system of only a few atoms. One might assume that for three atoms the vast body of quantum theory could provide more than a little insight. So as a simple but biologically important example, consider a molecule of water (H_2O). Water consists of two hydrogen nuclei with masses equal to m_p and electric charges equal to $+e$, an oxygen nucleus with a mass of $16m_p$ and a charge of $+8e$, and ten electrons with masses of $m_e = m_p/1836$ and charges of $-e$. We note that the total electric charge $(+e + e + 8e - 10e) = 0$, and the total valence $(+1 + 1 - 2) = 0$, which suggests to the theoretical chemist that the molecule is stable. As can be seen from Figure 2, a mixture of the oxygen's $2p$ states leads to an electron cloud that is shaped somewhat like a boomerang or a banana. Thus the configuration of water with lowest energy is of the form

$$\overset{-}{O}$$
$$H \ + \ H$$

where the angle between the hydrogens is found to be 104.5 degrees.

This angular configuration of a water molecule implies a concentration of negative charge on the oxygen and a corresponding concentration of positive charge near the hydrogen atoms. Although the entire water molecule remains electrically neutral, the displacement of internal charge makes it behave as the electric equivalent of a magnet. This configuration, together with the fact that water molecules can easily rotate as a single unit amid other water molecules, profoundly influences the electrical properties of liquid water. In the terms of electrical engineering, the molecule has a dielectric constant of about 80, which can be compared with values from 2 to 4 for ordinary materials like glass, wood, plastic, etc. The high dielectric constant, in turn, has decisive implications for the physical structure of a cell membrane, as we shall see in Chapter 4.

The fact that H_2O is not shaped like

H O H

is an example of a seemingly inconsequential result at one level of the scientific hierarchy that has crucial importance at higher levels.

In the Schrödinger equation for a water molecule, as it turns out, the wave function will depend on the position coordinates of the three nuclei and the ten electrons. And since approximations are required to study individual atoms beyond hydrogen and helium, they are even more necessary in this case, which is much more complicated.

One important approximation method was introduced in 1927 by Max Born and a young American postdoctoral student named Robert Oppenheimer, who was later to lead the atomic bomb project at Los Alamos. Their method is based directly on Schrödinger's work (1926), and takes advantage of the fact that the mass of an electron is smaller (by a factor of several thousand) than the mass of a typical atomic nucleus (Born and Oppenheimer, 1927). So from the perspective of classical physics, the picture is of rapidly moving electrons buzzing like bees among the ponderous nuclei of a molecule. The mathematical formulation of the Born-Oppenheimer approximation is sketched in Appendix C, but the gist of it is to determine the internal potential energy of a molecule as a function of the positions of all the atomic nuclei. This function, called the Born-Oppenheimer potential energy surface, is the grail of quantum chemistry.

For a molecule composed of N atoms, the Born-Oppenheimer function can be written as

$$U_0(x_1, y_1, z_1, x_2, y_2, z_2, \ldots, x_N, y_N, z_N),$$

where x_1 is the x-coordinate of the first atom, y_1 is the y-coordinate of the first atom, ..., and z_N is the z-coordinate of the Nth atom. The change in U_0 as each of these components $(x_1, y_1, \ldots, z_N)$ is varied gives the force in that particular direction. Thus a knowledge of U_0 together with Newton's second law (acceleration equals force divided by mass) allows one to make classical calculations of the ways that a chemical molecule can vibrate.

Having formulated the Born-Oppenheimer equations, however, the quantum chemist is faced with the problem of actually solving them. This, alas, calls for further approximations. Numerical techniques have been developed for the Born-Oppenheimer approximation using assumptions that are similar to those employed for multielectron atoms. Such methods are based on a *smörgåsbord* of approximate basis functions because—unlike Swedish food—none is particularly good. For small molecules, such as H_2O, these methods give results that are quantitatively useful. But as the number of atoms in the molecule increases, computational requirements become impractical, and errors are no longer under control.

In general, the potential function U_0 has the following qualitative character: i) As two atoms approach each other, U_0 increases because of electrostatic repulsion between the positively charged nuclei, and ii) As the atoms separate from each other, they cease to interact. U_0 thus approaches a constant value, and the interatomic forces approach zero. Between these two limits, U_0 has a minimum at some particular values of $x_1, y_1, \ldots, z_N$ that define the structure of the molecule.

Now we must ask what actually happens in the laboratory. If the molecule being studied has more than a few atoms, and the chemist is unable to determine U_0 from the equations of quantum theory, she does not give up and turn to another profession. It is then assumed that the nuclei interact only between pairs, and a *phenomenological* potential $\tilde{U}_0$ is constructed from experimental data as an approximation to U_0. (An example of such a phenomenological potential is described at the end of Appendix C.) Although this is an unhappy state of affairs from the perspective of a reductionist, a constructionist, like American physicist Philip Anderson (1972), rather expects it. One has moved up to the next level of the scientific hierarchy.

Chemists study the dependence upon time of molecular systems using two different methods. The first is called *molecular dynamics*. Using experimentally based approximations to the Born-Oppenheimer surface and commercially available computer codes that are based on Newton's laws of motion, it is now possible to calculate the motions of the atomic nuclei in a chemical molecule. At this level of approximation, the linear quantum theory of atomic structure has been transformed into a nonlinear theory of classical motion. These calculations, it turns out, *ignore* quantum effects—except as they are embodied in the Born-Oppenheimer potential—yet they remain difficult to carry out for large molecules. But they do give the chemist some idea of how a collection of atoms might wiggle and dance.

Although molecular dynamics calculations are extremely complex, chemical engineers and living organisms seldom, if ever, grapple with a single molecule in isolation. Typical reactions in beakers and test tubes involve very large numbers of molecules, so statistical equations can be written that are valid on the average. This is akin to actuarial calculations performed by insurance companies and to the polling studies of elections that are done by political organizations. The actuary has no idea when or how a particular person will die and the pollster cannot say how that same person will vote, but both make predictions that hold well for large numbers of people.

And so we come to the second major method chemists use to evaluate the time dependence of molecular systems: *chemical rate equations*. An example of a process that can be described this way is the burning of hydrogen (H_2) by oxygen (O_2) to obtain water (H_2O) and heat. If many molecules of hydrogen and oxygen are allowed to interact as they move about in space, the process, called nonlinear reaction and diffusion, is among the most challenging of modern applied mathematics. A seemingly simple example is the candle. The chemistry is by no means elementary: one must consider the wick, the oxygen, the heat, and the vaporization of solid wax into fuel for the flame. In Chapters 4 and 5, we shall consider nonlinear reaction and diffusion of an electrochemical potential along a nerve fiber and through a nerve cell. But for now we can say without qualification that the laws of both classical and quantum physics, important as they are on their own terms, on their own levels of the hierarchy, cannot be directly applied to account for the dynamics of a candle flame.

Indeed, as we further consider the glassblower and the florist, we must reexamine the degree to which the rules of each honorable craft are tightly bound to the other, or to which they are separate disciplines. Interestingly, the extent of a connection (or lack thereof) was seen as somewhat problematical even at the close of the 1920s. In 1929, the English physicist Paul Dirac (1902–1984) had expressed the opinion that:

> The underlying physical laws for the whole of chemistry are completely known, and the difficulty is only that the exact application of these laws leads to equations much too complicated to be soluble.

More than a half century later, the picture is about the same. Many basic facts of chemistry are embodied in Mendeleev's periodic table. But it is beyond the table, in the emergence of molecules and in the calculation of their dynamic behavior, that serious if not unsurmountable computational difficulties arise. Before assuming that chemistry can be derived from physics, one should consider several points.

Physics, first, cannot predict *all* of chemistry; this is precluded by the immense number of molecular combinations that can emerge from atomic physics. Almost seventy years—and several computer revolutions—after the publication of Schrödinger's great work, little of molecular dynamics in the real world is derived from physics. As we have seen, what has been achieved depends mainly on approximate computations in the context of the Born-Oppenheimer scheme for *small* molecules.

Such molecules are of importance to chemists but of little note to biologists, who traffic in molecules several thousand times as large. What's more, the errors in these computations are often unknown. Chemical rate equations, which describe the events of real interest to chemical engineers and biologists, are unrelated to Schrödinger's equation. The chemical engineer has no more need for quantum theory than the actuary has for geriatrics or the political pollster for behavioral psychology.

Indeed, for those who are uncommitted to the position that physics can explain everything, it seems that the rules of the glassblower cannot immediately assist the florist.

QUANTUM CONSCIOUSNESS

The loss of predictability between one level of the scientific hierarchy and another was memorably expressed by Erwin Schrödinger in 1935. He imagined the example of a cat, tightly sealed in a box with a fifty percent chance of being killed therein. Before the box is opened, Schrödinger pointed out, the conventional rules of quantum mechanics suggest that the state of the cat should be represented by a wave packet mixing the quantum representation of a live cat with that of a dead cat. This mixed wave function would—according to a standard application of the rules of quantum mechanics—collapse into a quantum representation of either a live cat or a dead cat when a conscious observer opens the box and peers in. Schrödinger viewed this thought experiment as a "burlesque example" of the difficulties that can arise when the rules of quantum mechanics are used at an inappropriate level of the scientific hierarchy. Following Schrödinger's lead, an estimate is presented in Appendix E of the time for the wave packet representing a dead cat to change into one representing a living cat. The result is a *very* long time: greater than

$$10^{10^{24}}$$

times the age of the universe!

Nonetheless, some physicists take Schrödinger's burlesque example quite seriously. Following ideas developed by Heisenberg and the Danish physicist Niels Bohr (1885–1962), as embodied in what has come to be called the *Copenhagen interpretation* of the meaning behind quantum mechanical calculations, the American physicist Henry Stapp (1993) has recently proposed a theory of consciousness that transcends classical Newtonian physics. According to this bold theory, the universe and everything in it is represented by a *universal wave function*. This wave function evolves in time as is required by Schrödinger's wave equation and determines the probabilities of various *events*. The occurrence of a particular event, on the other hand, is represented by the collapse of some part of the universal wave function. Just as the wave packet describing a cat that is both alive and dead is said to collapse into a wave packet representing one or the other condition when a conscious observer opens the box and looks in, the formation of an idea in the brain of William Shakespeare is viewed as the collapse of some aspect of the wave function that describes the activity of his brain.

In deference to cat lovers—among whom the present author is included—one need not perform Schrödinger's experiment on a living cat. The statuette of porcelain sitting on my desk will serve as well, or the lovely vase just finished by our glassblower-physicist. As he hands it to the florist-chemist, it slips through their fingers and falls toward the floor. Afraid for the vase, both shut their eyes and block their ears, but the glassblower knows from long experience (and careful calculations) that there is exactly a fifty percent chance for it to be smashed. He tells the florist of a quantum mechanical wave packet—describing both the broken and the unbroken vase—that will instantly collapse into one or the other when they look down at their feet. The florist sighs, rolls her eyes, and replies with the words of an anonymous chemist:

> I know your Schrödinger's equation. I have even used it, but I don't *believe* it.

References

P W Anderson. More is different: Broken symmetry and the nature of the hierarchical structure of science. *Science*, 177:393–396, 1972.

J J Bollinger, D J Heinzen, W M Itano, S L Gilbert, and D J Wineland. Test of the linearity of quantum mechanics by rf spectroscopy of the ^{9}Be$^+$ ground state. *Phys. Rev. Lett.*, 63:1031–1034, 1989.

M Born and J R Oppenheimer. Zur Quantentheorie der Molekeln. *Annal. Physik*, 84:457–484, 1927.

L de Broglie. *Non-linear wave mechanics: A causal interpretation.* Elsevier Publishing Co., Amsterdam, 1960.

P A M Dirac. Quantum mechanics of many electron systems. *Proc. Royal Soc. London* 126A:714–733, 1929.

W M Elsasser. *Atom and organism: A new approach to theoretical biology.* Princeton University Press, Princeton, 1966.

W M Elsasser. Acausal phenomena in physics and biology: A case for reconstruction. *Am. Scientist*, 57:502–516, 1969.

T Ferris, editor. *The world treasury of physics, astronomy, and mathematics.* Little, Brown and Co., Boston, 1991, p. 808.

A R von Hippel. *Molecular science and molecular engineering.* MIT Press/Wiley, New York, 1959.

W Moore. *Schrödinger: Life and thought.* Cambridge University Press, Cambridge, 1989.

E Schrödinger. Quantisierung als Eigenwertproblem. *Ann. Physik*, 79:361–376, 1926.

E Schrödinger. Die gegenwärtige Situation der Quantenmechanik. *Naturwissenschaften*, 23:807–812, 823–828, and 844–849, 1935.

J C Slater. *Quantum theory of matter.* McGraw-Hill, New York, 1951.

J C Slater. *Quantum theory of atomic structure.* McGraw-Hill, New York, 1960.

H P Stapp. *Mind, matter, and quantum mechanics.* Springer-Verlag, Berlin, 1993.

S Weinberg. Precision test of quantum mechanics. *Phys. Rev. Lett.*, 62:485–488, 1989.

The Chemistry of Life

Biology does not deny chemistry,
though chemistry is inadequate
to explain biological phenomena.

Ruth Benedict

I t is the end of a warm summer day. The sun has set and
night is falling over the countryside. In a meadow—now cool—
thousands of fireflies are climbing awkwardly up the blades of
grass in preparation for their nocturnal adventures. Soon they spread their
wings and fill the air with countless flashing lights that form pattern after
pattern in the darkness: an endless display of luminescent intercourse. Sitting
in the shadows and watching, we wonder at the source of this light. Clearly
it is a far more subtle phenomenon than the light from the glassblower's
oven that we looked at in the previous chapter. Where does the light come
from? What turns it on and off? If we could answer such questions, would
we know anything about the nature of consciousness?

It turns out that there *is* a connection between the firefly's strange light
and our mysterious sense of consciousness. We learn this from the anesthesiologists who have become familiar with a dozen or so chemical substances
that have greater or lesser power to block awareness for a limited period
of time. These anesthetic agents differ widely in their chemical composition
and in their effectiveness, but surprisingly, the ability of a particular agent
to suppress consciousness equals its ability to reduce the light from a firefly
(Adey et al., 1975; Franks and Lieb, 1982).

BIOMOLECULAR STRUCTURES

The firefly's light is emitted by luciferase, a large biological molecule classified as a protein. The way that luciferase works depends upon its atomic
structure, and this structure, in turn, is related to a genetic code that resides
in the nuclei of the firefly's cells. The nucleus contains small fibers resem-
bling pipe cleaners called chromosomes. Human beings have 23 pairs of
chromosomes; fireflies possess a different number as allotted by evolution.
Each chromosome, in turn, contains strands of the marvelous double helix

27

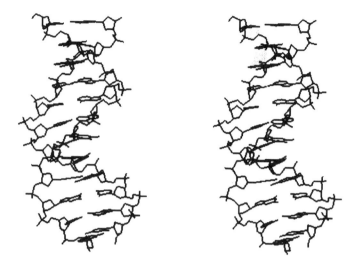

Figure 5. *A stereoscopic view of a short section of DNA. To see the three-dimensional structure, hold the figure a few inches from your nose and let your eyes relax. Notice the two helical backbone structures to which the (inward directed) base pairs are attached.*

of deoxyribonucleic acid (or DNA) first identified, in 1953, by Francis Crick and James Watson using the x-ray data of Rosalind Franklin and Maurice Wilkins.

We begin with a review of the structure of the double helix because, as a relatively well-understood natural structure, it will help us comprehend more elaborate cellular machinery that has yet to be fully understood. (The atomic structure of DNA is shown in the stereographic plot of Figure 5, where the line segments indicate distances between centers of valence-bonded atoms.) The two helices, like a ladder, can be imagined with rails running lengthwise and rungs running crosswise. The rails are *backbones* of sugars and phosphates and, like the steel girders supporting a skyscraper, are now more or less taken for granted by biochemists studying the architecture of DNA.

But the rungs of the helix are of paramount interest. The rungs consist of four bases. Thanks to complementary shapes and molecular forces that guide them together, these bases are always paired off in a highly repetitive fashion. Adenine (A) is always joined with thymine (T); cytosine (C), in turn, is always coupled with guanine (G). Thus the series of half-rungs that might be found along one rail of the helix, say

–A–C–T–C–G–A–G–T–C–A–T–,

could only be paired with the half-rungs of

–T–G–A–G–C–T–C–A–G–T–A–.

As it turns out, one strand of DNA is actually the template to create a complementary strand. And this complementary strand, at a later stage, letter by letter, itself serves as a template for stamping out protein molecules like luciferase. Cells manufacture a wide variety of proteins after consulting their internal blueprints, the DNA, to find the appropriate structures.

Proteins are constructed of amino acids, each of which has the general formula:

$$RCH(NH_2)COOH,$$

where "R" indicates a *residue* that can be one of twenty molecular groups. By giving up a water molecule (H_2O) at each connection, amino acids can form a chain, and this chain is a protein. Thus, each amino acid in this chain can be one of twenty different sorts, some positively charged, some negatively charged, and some neutral. The order of amino acids along the chain is called the *primary structure* of the protein. This primary structure is determined by a DNA code.

Since one base pair in a DNA template can be chosen in four different ways, two can be chosen in four times four (4^2) or sixteen ways, and three in four times four times four (4^3) or sixty-four ways. Thus a triplet of DNA bases is easily able to specify each of the twenty amino acids.

How many proteins can there be? Suppose for a moment that a certain protein consists of only two amino acids linked together. There are twenty ways to choose the first of these, and twenty for the second. Then, for N equal to two, there are 20^2 or 400 possible protein structures. In the general case of N amino acids, therefore, there are 20^N possible proteins. For a normal protein, this is a very large number. Setting 20^N equal to the immense number $\Im$ (which was defined in Chapter 2 as 10^{110}) tells us that the number of proteins is immense for N greater than 85. Since a typical protein has 200 or more amino acids in sequence, it is clear that the number of possible protein molecules is immense.

In more graphic terms, imagine that one were to try to make just one single molecule of every possible protein. The entire mass of the universe would not suffice for the task, so in the future course of evolution there will always be many interesting protein molecules that have never been employed by a biological organism. Thus do immense numbers limit one's ability to predict the facts of biochemistry from the laws of physics.

Now, how many different strands of DNA can be constructed? Considering the bases that are attached to a single strand, the first may be chosen from four possibilities: A, T, C, or G. Likewise for the second, and so on, up to a total of N base pairs for which there are 4^N choices. Setting 4^N equal to the immense number $\Im$, we find that the number of DNA strands with more than 184 base pairs is immense. Typical DNA strands in mammals, which have on the order of a hundred million base pairs, can be arranged in something like

$$10^{10^8}$$

different ways. All the individual organisms that have ever lived have tried but a minute fraction of these immense genetic possibilities.

At this point we must digress about proteins, because their roles throughout the body and brain are so complex and so varied.

Our hair and skin are made of a protein called keratin, which is also the crucial ingredient in biological materials like feathers and fingernails, hooves and horns. Another protein, collagen, is the stuff of connective tissues such as tendons, bones, and cartilage. Fibroin is the silk of cocoons and spider webs. Sclerolin forms the external skeletons of insects such as the firefly. Such proteins are *structural*, lending shape and form to life. These are only a few of the more notable examples.

Still other proteins are more *functional*, taking crucial roles in other processes within cells. Insulin is a well-known protein that regulates the metabolism of glucose, a natural sugar. Rhodopsin, in the retina of the eye, converts incoming light to ionic signals in the optic nerve. Myosin and actin provide the forces in vertebrate muscle cells. Dynein is an energy-inducing component of the cytoskeleton. Hemoglobin carries oxygen in the blood, and myoglobin (a close cousin) carries oxygen in muscle. Almost all enzymes—catalysts that speed biochemical reactions but are not consumed by them—are functional proteins. So are the transmembrane sodium, potassium, and calcium channels in nerve membranes, which support the basic electrochemical activity of the brain. And so on through what we know of an immense number of possible structures.

Let's take a look at myoglobin, which carries oxygen in our muscles. It has the chemical formula

$$C_{738}H_{1166}FeN_{203}O_{208}S_2,$$

indicating that it is composed of 738 carbon atoms, 1166 hydrogens, one iron, 203 nitrogens, 208 oxygens, and two sulfurs. The three-dimensional structure (or *tertiary structure*) of myoglobin is shown in Figure 6, where the little spheres indicate the quantum mechanical wave functions that were the subject of the previous chapter. The iron atom forms part of an active site, called the *heme group*, that binds oxygen deep inside. Hans Frauenfelder, an experimental biophysicist who has spent the past twenty years trying to understand the dynamics of myoglobin, calls it the "hydrogen atom of biochemistry."

Below its tertiary structure one finds *secondary structures*. The darker region in the stereo images of myoglobin in Figure 6 has a coiled shape, called the *alpha-helix*, which is shown in greater detail in Figure 7. In this arrangement, the alpha-helix is formed by longitudinal hydrogen bonds between the oxygen of one amino acid and the nitrogen of another, three amino acids farther along the chain. Proteins like keratin (hair) and myosin (in muscle) are almost entirely alpha-helical, but even a globular protein, like myoglobin, usually contains several alpha-helical segments. Another example of

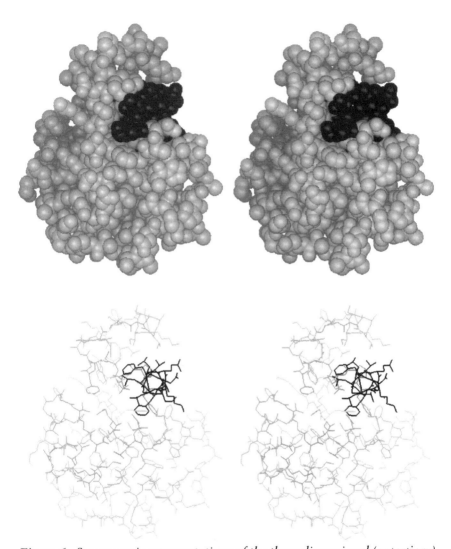

Figure 6. *Stereoscopic representations of the three-dimensional (or* tertiary*) structure of myoglobin. The darker shading indicates an alpha-helical region of* secondary *structure. A space-filling image in which the electronic wave functions of the atoms are indicated (upper). The line segments indicate the distances between the atomic centers so the entire three-dimensional structure can be seen (lower).*

secondary structure in proteins is the *beta-sheet*, in which hydrogen bonds take hold between adjacent segments of primary strands. The beta-sheet resembles a Venetian blind, each slat representing a primary strand.

It is important for us to understand the secondary structures of proteins because they provide avenues of communication at the most basic biological level (Scott, 1990). Transmembrane proteins, which carry ions into and out

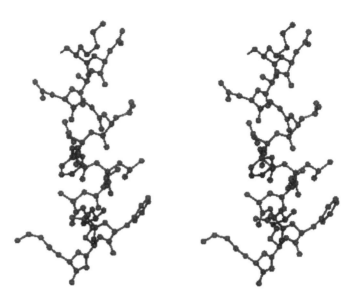

Figure 7. A stereoscopic view of the section of alpha-helix in myoglobin that is emphasized in Figure 6.

of a cell, often consist of about seven alpha-helix segments in the form of an *ionic channel*. If these channels become blocked in nerve cells, consciousness may disappear.

Many functional proteins have an active site (like the heme pocket in myoglobin) at which the action occurs. Although the three-dimensional structure of luciferase has not yet been determined, there is evidence that its switch is located at an active site. Similarly, if you ever need surgery, your anesthesiologist will provide your body with appropriate chemical agents that manage to block the active sites of proteins that are related—somehow—to consciousness without disturbing those that regulate your breathing and the beating of your heart. This miracle of modern medicine is not yet understood, and it is the tip of an iceberg (or the head of a camel) called *psychoneuropharmacology*. We know, of course, that changes in the glucose level of the blood or in its ionic composition can cause hunger or thirst, as sex hormones or adrenaline can induce lust or courage. Beyond this, an ever-growing collection of artificial and natural drugs complements the traditional molecules of caffeine, ethanol, and ether to act on the control centers of key brain proteins. Such molecules—drawn from an immense number of chemical possibilities—cause ever more strange alterations in our states of consciousness. Wakefulness and sleep, elation and depression, ecstasy and paranoia can all be so induced.

At the social level, economic relationships are disrupted and neighborhoods are threatened or destroyed by the various actions of certain chemical molecules on the active sites of certain proteins in certain neurons that operate in certain regions of the brains of certain people who occupy certain positions in the vast fabric of human culture. Perhaps this field should be called *ethnopsychoneuropharmacology*. In any event, the investigation of protein interactions with chemical molecules cannot be overlooked by the serious student of consciousness.

Proteins are delicate molecules. In order to function properly, a protein must maintain its tertiary structure, but this can change. If the tertiary structure is altered through the action of heat (think of frying egg white) or of a chemical (try putting lemon juice into milk), the protein is said to become *denatured*. Although the primary structure is unchanged, the secondary and tertiary structures of a denatured protein are altered enough so that the protein can no longer function. Anyone who has spent time in the kitchen knows that a steak is easier to cut and digest *after* it has been denatured than before.

Since the 1960s, the tertiary structures of several hundred proteins have been determined by the method of x-ray crystallography. Computers are used to calculate and record the atomic coordinates of every atom in the molecule, and these coordinates are available electronically from the Protein Data Bank of the Department of Chemistry, Brookhaven National Laboratory, Upton, L.I., New York. It has become relatively easy for researchers to search for existing structures or submit the atomic coordinates of new structures that they have identified.

There is one unfortunate obstacle to this important progress. To determine the structure from x-ray diffraction, it is necessary to form a crystal that contains many examples of the protein. Although difficult, this is possible for water-soluble proteins. But it has not been done for the lipid-soluble proteins that one finds in the cell membranes. The lipid-soluble proteins, alas, are vitally important because they mediate the flows of ions into and out of cells, which are basic both to life and to consciousness.

Now the biochemist would like very much to be able to predict the tertiary structure of a protein (refer again to Figure 6) from its primary structure (the series of amino acids). The quest for this prediction is known as the protein-folding problem, and it remains famously unsolved, despite massive supercomputers, nearly a century of quantum physics, and nearly three centuries of competent chemistry. If scientists *could* predict the tertiary structure from the primary structure, any molecule imaginable could be created. That is because biotechnology now allows scientists to snip and edit and paste the primary structure at will.

But for the moment, any correlation between a new sequence of amino acids and a new tertiary structure is largely unpredictable. This itself is surprising, given that chemists and physicists know all of the elements in the realm of biochemistry. In small molecules they can predict some molecular interactions reasonably well. It is in the area of biological molecules that they are far from understanding exactly how it is that a particular genetic sequence of A, C, T, and G can give rise to a protein in a whipping flagellum on a simple pond organism. This seemingly simple problem of energy minimization is maddeningly difficult because the ground state of a protein molecule is not unique (van Gunsteren, et al., 1993; Mielczarek, et al., 1993). Most of the residues can assume several positions (say n) without greatly changing the energy of the structure, so a protein of N amino acids will have about n^N ground states of approximately the same energy. With $n = 4$ and $N > 184$, the number of ground states is immense. Since all of these ground states must be compared before deciding which has the lowest energy, a direct approach to solving the folding problem is not feasible.

In addition to the artificial crystals of protein that are prepared for x-ray structure determinations, some proteins naturally take a shape that is approximately that of a crystal. For our purposes in this book, the most fascinating example is tubulin. It forms cylinders called microtubules, shown in Figure 8. The roles of tubulin, though not as widely appreciated as those of myoglobin or keratin, are equally fascinating. They include:

Flagella and cilia. We all have seen one-celled organisms swimming around under the microscope. The molecular engine driving such locomotion is a biochemical complex built of tubulin, dynein, and other intracellular energy sources. Just as myosin and actin interact to provide the force of vertebrate muscle, microtubules with tubulin and dynein provide the biological motors for flagella and cilia. These structures, crystalline bundles of tubulin arranged in nine supercylinders, project from the cell surface. The result for a single cell is propulsion and for a fixed cell is the movement of nearby liquids past the cell (Murase, 1992).

Centrioles. Microtubules also help cells divide. The same cylindrical bundles of tubulin are the organizing lattice for cell division, keeping track of the structure of the parent cell's nucleus as it separates to initiate mitosis and establish the architecture of the daughter cells. The underlying dynamics of this fascinating and important phenomenon are poorly understood.

The cytoskeleton. As is shown in Figure 8, the inside of any cell with a nucleus is filled with a complex scaffolding network called the cytoskeleton. Originally, the microtubules of tubulin protein were believed to provide structural support for the cell.

Nowadays some suggest that the cytoskeleton may also provide the hardware for an intracellular computer (Hameroff and Watt, 1982), an idea that is not as far-fetched as it might sound. Single-celled organisms (such as the amoeba and paramecium) need something of this sort as they forage and feed, conjugate and multiply in response to the happenstance events of their little lives. If we suppose that individual cells have internal computers, it becomes interesting to ask how such computers might be related to the global consciousness of many-celled organisms like ourselves. It has been suggested that cytoskeletal dynamics may be quantum mechanical in nature and thus a component in the mechanism of quantum consciousness that was mentioned in the previous chapter. We shall return to this theme in later chapters.

What are the facts? Microscopic movies of extended nerve fibers (called *axons*) clearly show the bidirectional transport of various materials, prob-

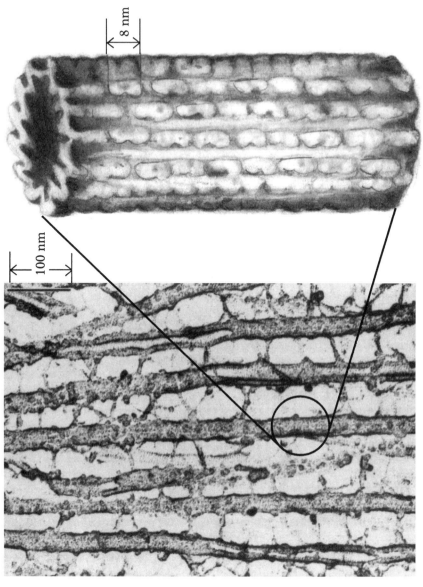

Figure 8. Electron micrograph of a neuronal cytoskeleton that has been expanded to show its structure as a protein crystal. One nanometer (nm) equals 10^{-9} meter or 10 Å. (Constructed from Hirokawa, 1991, and Amos and Amos, 1991.)

35

ably nutrients and wastes. The image is rather like that of activity in the corridor of a busy warehouse. Since cytoskeletal structures pervade the cell and are known to move (recall the behavior of cilia and flagella), they are probably implicated in this activity. And a warehouse cannot function without some sort of information system to keep track of the comings and goings of the parcels.

All these elements of life emerge from the atoms of physics and the simple molecules of inorganic chemistry—under the influence of nonlinear Born-Oppenheimer potential energy functions—to form a new level of complexity. This new level should be approached with a sense of wonder: it is more intricate than even the most elaborate scheme concocted by the most creative scientist. Why? Because a scientist, at the age of 60, has had perhaps 40 years to work, but Nature has been constructing the main components of biochemistry for several billion years. These myriad reactions—each tailored to a specific need—form a network of dynamic activity that stuns the imagination, leaving it humble before the biochemical reality of a single cell, let alone an entire organism.

BIOMOLECULAR DYNAMICS

To this point, we have been concerned primarily with the *structures* of biochemical molecules, but life involves motion and change. Thus it is also of interest to consider the science of biomolecular dynamics (McCammon and Harvey, 1987; van Gunsteren et al., 1993; Grønbech-Jensen and Doriach, 1994). The basis for any such study is the Born-Oppenheimer potential energy function, which was discussed in the previous chapter and defined in Appendix C. This function (called U_0) gives the forces between the atoms of a molecule in terms of their distances of separation.

Since it is not possible to compute U_0 for molecules that are composed of more than a few atoms, chemists turn to an experimentally based approximation to the Born-Oppenheimer potential energy surface, such as $\tilde{U}_0$ given at the end of Appendix C. Then one uses Newton's second law (force equals mass times acceleration) together with $\tilde{U}_0$ to obtain dynamic equations for the atomic coordinates. In addition to providing insight into the nature of protein motion, molecular dynamics calculations are also useful for refining x-ray structure data by letting the system evolve toward its energetic minimum.

How reliable are such calculations? In the course of their recent survey of numerical techniques for biomolecular dynamics, Andrew McCammon and Stephen Harvey (1987) sound the following note of caution:

> Given the wide range of problems that can be attacked by molecular dynamics, it is worthwhile to examine some of the limitations of this technique. Most of these grow out of the fact that the

molecular models are so detailed (typically including at least all heavy atoms and frequently many or all hydrogens) and the time step is necessarily so short (a fraction of the period of the highest frequency motion, and hence on the order of $[10^{-15}$ second]) that only relatively small systems and short time scales are accessible to simulations. Total molecular weights (including solvent) are typically on the order of 25,000 or less, and the total period covered is typically a fraction of a nanosecond. With the availability of supercomputers, array processors, and multiprocessors, these values may be pushed up by one or more orders of magnitude, but the accessible time scales will still be far shorter than those of most biologically interesting processes.

Despite such caveats, the curious biochemist is tempted to use the tools of biomolecular dynamics to uncover some aspects of protein motion. In the linear approximation, a protein can exhibit several different motions at particular frequencies—like the individual notes of a violin—but the general behavior is *nonlinear* and therefore considerably more complicated. From a wide variety of experimental and theoretical studies on myoglobin, Frauenfelder and his colleagues (1991) have shown that a typical protein at normal physiological temperature is far from the static structure revealed by x-ray crystallography and displayed in Figure 6. Instead it is "screaming and kicking" in a "bewildering variety" of motions that have been revealed by fluorescence spectroscopy, nuclear magnetic resonance, hydrogen exchange, and Raman scattering. This motion of a tiny molecule of myoglobin is organized in a hierarchy of dynamic levels, much—as we shall see in Chapter 6—like the brain itself.

The physicist's and chemist's predicament only worsens when they turn to the problem of understanding *energy* in the realm of biochemistry. Let's examine one reaction in particular. The hydrolysis of adenosine triphosphate (ATP) into adenosine diphosphate (ADP) releases—under normal physiological conditions—about 10 kcal/mol or 0.42 electron-volts of free energy. This energy serves as a source for biological activity (Fox, 1982); for example, ATP switches on the light from the firefly's luciferase. One can imagine it as a cellular fuel tank, an ATP battery, that naturally runs down. As it does, it becomes necessary to convert ADP back into ATP. But how?

In cells with nuclei, the process takes place in the mitochondria and is called *oxidative phosphorylation*. Mitochondria are subcellular organelles that contain intricately folded inner passages. They may have evolved from separate organisms; a cartoon of one of them is sketched in Figure 9. In this sketch, the membrane is a double layer of fatty molecules, all aligned so that their charged, water-friendly head groups are directed outward toward the water, and their uncharged hydrophobic tails are shielded from water in the interior of the membrane. Energetically, this is a stable configuration: the high dielectric constant of water (about 80, as was noted in the previous chapter) reduces the electric field energy near the head groups.

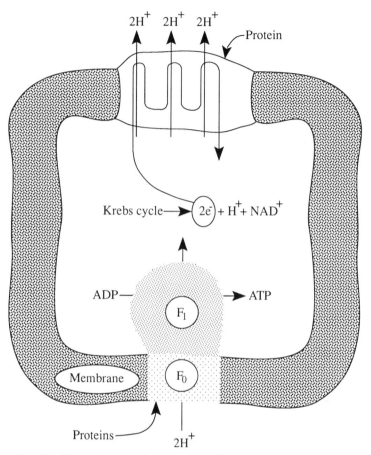

Figure 9. *Simplified sketch of a mitochondrion. Only a single membrane protein of each type is indicated.*

The mitochondrion's pure, double-layered fatty barrier is an excellent insulator. Thus, ionic currents through the membrane are mediated entirely by proteins embedded in the membrane. Within the mitochondrion, a cycle of chemical transformations called the citric acid (or Krebs) cycle breaks down the carbon chain of glucose to carbon dioxide. Broadly viewed, the effect of the Krebs cycle is fairly straightforward. Fats and carbohydrates are digested—by appropriate enzymes—into acetic acid (CH_3COOH), which is oxidized to release energy. It turns out that a single molecule of oxaloacetate is used over and over again to oxidize many acetate residues. In the process, this reaction produces nicotinamide adenine dinucleotide (NAD^+) plus a proton (H^+) and two electrons ($2e^-$). These two electrons make three trips across a region of intrinsic membrane protein, carrying two protons from inside the mitochondrion to the outside on each pass. The result of this process is twofold: First, the pH inside is increased by about 1.4 with respect to the outside; thus the inside becomes less acidic. Second, the membrane potential is increased on the outside by about 140 millivolts.

The upshot of all this is that the membrane-enclosed volume of an active mitochondrion is negatively charged inside and hydrogen ions are at a higher concentration outside. Both of these effects drive hydrogen ions from the outside to the inside. Energy from the glucose has been transformed into the tension experienced by the hydrogen ions.

What next? The inward pressure of hydrogen ions causes a flow through the membrane proteins labeled F_0 and F_1 in Figure 9. This inward proton current—somehow—manages to convert ADP back into ATP.

Biochemists have mapped out hundreds of such reactions in living organisms, but many more remain to be studied. Once again we must allow our jaws to drop for a moment at the awesome complexity of a living organism, comprising *thousands* of such reactions, each exquisitely tailored to its biological needs. How—one asks—might all of this dynamic activity be organized?

This question was addressed about fifteen years ago by Manfred Eigen and Peter Schuster in an important book entitled *The hypercycle: A principle of natural self-organization* (1979). In this book they showed that life can be viewed as the activity of an interrelated hierarchy of cyclic reaction networks, which they call a *hypercycle*.

A hypercycle is a cycle of cycles of cycles. One example of a basic catalytic cycle is the citric acid cycle, which uses a molecule of oxaloacetate over and over again to extract the energy from acetic acid. We have just seen how the citric acid cycle drives a variety of higher-level cycles (like the flashing of the firefly's light) that require the consumption of energy stored in ATP. At a higher level of organization, several such basic catalytic cycles are organized into an autocatalytic cycle, which is able to instruct its own reproduction. An example of this is a complex protein-DNA structure that can reproduce itself. Several of these autocatalyst cycles are then organized at a higher level into a *catalytic hypercycle*—such as a virus—which stands at the threshold of life because it can evolve into ever more efficient forms. According to Eigen and Schuster, the basic structure of life is organized in the following way:

Catalytic hypercycle
↑ ↓
Autocatalytic cycle
↑ ↓
Basic catalytic cycle

Can a similar picture be applied to describe the nature of consciousness?

IS LIFE BASED ON THE LAWS OF PHYSICS?

What role does physics play in the chemistry of life? At the level of the biodynamic calculations surveyed by McCammon and Harvey, quantum theory is no longer in the picture. Quantum effects are neglected whenever one uses Newton's second law of motion. What about classical physics? Does it make sense to argue that the operation of a biological hypercycle exemplifies only the laws of classical physics? Isn't this rather like saying that a television set obeys only the laws of electricity? Or an automobile obeys only the laws of mechanics?

Another way to pose such a question has been suggested by the American physicist Murray Gell-Mann (1994), who wonders whether the nature of biochemistry is uniquely determined by the laws of physics. Might life on another planet—while obeying the same fundamental laws of physics and chemistry—evolve in a different way? Could life out there employ the chemistry of silicon, say, or germanium instead of carbon? Or is all life throughout the universe based on the molecules of DNA (as shown in Figure 5) and on proteins constructed from the twenty amino acids with which we earthlings are familiar? Gell-Mann wonders if "biochemistry depends mainly on asking the right questions about physics or also depends in an important way on history?"

Uncertainties about the applicability of physical laws to biology are nothing new and in fact predate by decades our modern knowledge of molecular biology. As long ago as 1943, Erwin Schrödinger presented three public lectures at Trinity College in Dublin, and the title he chose was "What is life?" In these lectures he was among the first to state in clear language the notion that the genetic message is written in a molecular code. His lectures were published the next year in a small volume that influenced developments in molecular biology in many ways, not the least of which was by attracting both James Watson and Francis Crick to the field (Moore, 1989). The last chapter of his seminal book asks "Is life based on the laws of physics?" Schrödinger's conclusion was:

> From all we have learnt about the structure of living matter, we must be prepared to find it working in a manner that cannot be reduced to the ordinary laws of physics. And that is not on the ground that there is any "new force" or what not, directing the behaviour of the single atoms within a living organism, but because the construction is different from anything we have yet tested in the physical laboratory.

References

L A Amos and W B Amos. *Models of the cytoskeleton.* Guilford Press, New York, 1991.

G Adey, B Wardly-Smith, and D White. Mechanism of inhibition of bacterial luciferase by anesthetics. *Life Sciences*, 17:1849–1854, 1975.

R Benedict. *Patterns of culture*. First published 1934, republished by Houghton, Mifflin, Boston, 1989.

M Eigen and P Schuster. *The hypercycle: A principle of natural self-organization*. Springer-Verlag, Berlin, 1979.

R F Fox. *Biological energy transduction: The uroboros*. John Wiley & Sons, New York, 1982.

N P Franks and W R Lieb. Molecular mechanisms of general anesthesia. *Nature*, 300:487–493, 1982.

H Frauenfelder, S G Sligar, and P G Wolynes. The energy landscapes and motions of protein. *Science*, 254:1598–1603, 1991.

M Gell-Mann. *The quark and the jaguar*. W.H. Freeman and Company, New York, 1994.

N Grønbech-Jensen and S Doniach. Long-time overdamped Langevin dynamics of molecular chains. *J. Comp. Chem.*, 15:997–1012, 1994.

W F van Gunsteren, P K Weiner, and A J Wilkinson, eds. *Computer simulation of biomolecular systems*. ESCOM Science Publishers, Leiden, 1993.

S R Hameroff and R C Watt. Information processing in microtubules. *J. Theor. Biol.*, 98:549–561, 1982.

N Hirokawa. Molecular architecture and dynamics of the neuronal cytoskeleton. In *The neuronal cytoskeleton*, RD Burgoyne, ed., Wiley, New York, 1991.

J A McCammon and S C Harvey. *Dynamics of proteins and nucleic acids*. Cambridge University Press, Cambridge, 1987.

E V Mielczarek, E Greenbaum, and R S Knox. *Biological physics: Key papers in physics*. American Institute of Physics, New York, 1993.

W Moore. *Schrödinger: Life and thought*. Cambridge University Press, Cambridge, 1989.

M Murase. *The dynamics of cellular motility*. John Wiley & Sons, Chichester, UK, 1992.

E Schrödinger. *What is life?* Cambridge University Press, Cambridge, 1944 (republished 1967).

A C Scott. The solitary wave: An enduring pulse first seen in water may convey energy in the cell. *Sciences*, 30:28–35, 1990.

The Nerve Afire

Companion of my griefs! thy sinking frame
Had often drooped, and then erect again
With shews of health had mocked forebodings dark;
Watching the changes of that quivering spark,
I feared and hoped and dared to trust at length,
Thy very weakness was my tower of strength.

Mary Wollstonecraft Shelley

S cientists and poets can agree that consciousness is like a bonfire—
a conflagration in which the mind blazes and retreats, only to
erupt again. Such flames, as we shall see in the next three chapters, burn within a hundred billion individual nerve cells in the human brain.
Because this fire in its native setting is not to be carelessly approached or easily manipulated, we must first examine the vital essence of the neurological
world: a single spark from one neuron.

The number of individual sparks triggered by the reading of this sen-
tence at this moment—the neural impulses lighted by this book—cannot
be calculated. But we *do* know enough about the smallest indivisible sub-
unit of consciousness to begin to appreciate the dimensions and beauty and
complexity of the larger blaze.

The spark of one nerve impulse begins with a stimulus. This can take any
number of forms. In the case of a lecturing professor, the stimulus may be
a sharp tap on the blackboard: students respond by paying closer attention.
In the realm of living creatures as they evolved from single-celled organisms,
the stimulus was often chemical in nature. The water near a squid, say,
bore molecular hints of prey, or the slightest scent of a predator. It would
behoove the squid to be able to detect a change in its surroundings and
respond quickly. Squids able to detect such hints survived better than those
unable to do so. And so evolution over eons shaped a system of exquisite
sensitivity. Its culmination in the visual realm may be the eye of the falcon;
in the olfactory realm, the chemoreceptors of a shark. But in all cases the
underlying mechanism—the neuronal spark—is the same.

The cellular equipment, too, is surprisingly similar across the animal
kingdom. By inspecting the images of neurons that we have all seen in
books and magazines, we can detect a central body in every one. Spiky
tendrils protrude out from this body. Most of the craggy branches are called
dendrites, from the Latin word for tree. Dendrites serve as the in-boxes

of the neuron's informational office, receiving biological signals within the body. A single neuron may have many dendrites with thousands of input terminals, but it generally has *one* central shaft called an axon, with its own appropriate etymological roots in the Greek word for axle. The axon is the out-box of the neuron's computational apparatus: it emits signals received by its dendrites nearby, just as the branches of a tree send nutrients into a taproot. No computer is composed of circuits so elegant.

In the realm of biology, one hardly need elaborate, there are no rotating electrical generators, no dynamos to power the cells. But there is nevertheless an energy source of equivalent potency. Biochemical processes, briefly described in the previous chapter, fuel the neuronal fires. The capacity for work comes from a small difference in the electric potential within and outside the cell. Neurons, like all cells, have semipermeable membranes that selectively admit a variety of ions, sodium and potassium ions being the main units of their currency.

But a neural membrane is distinctive in that its permeability for a particular ion (say sodium) is sensitive to the electrical voltage across the membrane. As ions rush in and out of the neuron, they carry electrical charge, which, in turn, alters the membrane voltage and leads to the formation of a *nerve impulse* as shown in Figure 10. This impulse travels up the trunk of the axonal tree and out to its most distant branches, carrying an item of information, a bit of mental data, to the input terminals of other nerves or of muscles.

Since the year 2600 B.C., when Egyptian records first described the electric catfish, the relationship between life and electricity has been a subject of intense interest and often controversial speculation. At the dawn of the nineteenth century, this interest blossomed into a variety of scientific experiments on the electrical stimulation of living tissue, and in the mid-nineteenth century, the German physicist and physiologist Hermann Helmholtz (1821–1894) first measured the velocity of nerve pulse propagation (Helmholtz, 1850). He did so despite the fact that his father, a philosopher, had advised him not to bother with the experiment. Since a volitional act is identical to the will that directs it, the elder Helmholtz reasoned, it would be a waste of time to try to observe a delay between the application of an electrical current and its effect. If one orders one's finger to move, many suppose, there is no notable delay.

Nonetheless young Helmholtz proceeded with his measurements, supplying electricity to the sciatic nerve of a frog and waiting for the resulting contraction in its leg muscle. He did observe a delay: from the moment of the application of the electricity, a few fractions of a second passed before the frog's muscle moved. Luigi Galvani (1737–1798) and others had performed similar experiments since the late eighteenth century, but it fell to Helmholtz to design and build an elegant experimental apparatus that enabled him to calculate the speed of propagation on a frog's sciatic nerve as about 2700 centimeters per second, which is close to the currently accepted value. Throughout the rest of the nineteenth century, controversy simmered over the cause of this delay.

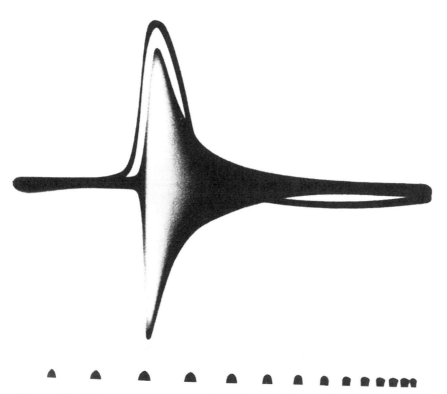

Figure 10. *Direct measurements of the increase in membrane ionic permeability (band) and transmembrane voltage (line) on the giant axon of the squid. The time increases from left to right and the marks indicate intervals of one millisecond. (Courtesy of Kenneth Cole.)*

Some scientists insisted that an animistic biological principle—some sort of electric soul—must be at work. (As we shall see in Chapter 7, not a few philosophers today hold more contemporary conceptions of the same idea with respect to the concept of consciousness.) Helmholtz himself found that approach unsatisfactory, but because he was not familiar with the phenomenon of nonlinear diffusion, he proposed that the nervous impulse was a pulse of cytoplasmic matter flowing through the nerve. This oversight was uncharacteristic. Helmholtz, after all, was among the leading mathematical physicists of his day. Among other achievements, he described the structure of vortices in liquid and, astoundingly, formulated the law of conservation of energy. Yet the notion of a steady flame as a model for the nerve impulse did not occur to him. Sometimes the simplest ideas are the most obscure.

NERVE MEMBRANE DYNAMICS

The squid, as it happens, possesses exceptionally interesting neural equipment, including two unusually large nerves, each about the size and texture of a piece of spaghetti, running along its back. Such large nerves, called giant axons, evolved to enable the little lady to escape quickly from a predator because large-diameter nerves carry signals at a greater speed than those of smaller diameter. (To further confuse the predator, she leaves behind a pseudomorphic cloud of tasty ink, which also makes a toothsome spaghetti sauce.) First reported to *be* a nerve in 1909, the significance of the giant axon was forgotten until its rediscovery in 1933 (Arnold et al., 1974). In the ensuing sixty years, it has been of intense scientific interest because one can do more precise experiments on a big nerve than on a little one.

The reader may wonder why we don't all have fat nerves to help us get a head start on our predators or our prey, as the case may be. The answer is that vertebrate species have evolved a means to speed up their nerve signals *without* increasing the diameter of the nerve fiber. This remarkable invention involves the wrapping of an insulating sheath—called *myelin*—about a small nerve fiber, which limits switching action of the nerve to widely spaced active nodes. The signal then jumps from one active node to the next, rather like the propagation of information about the presence of pirates along the Amalfi Coast during the Middle Ages. A fire built atop one watch tower was seen at the next tower, whereupon its fire would be lit to signal the next tower, and so on. Clearly, the speed of the nerve signal (or pirate warning) depends primarily upon the spacing between the active nodes (or watch towers), and this spacing can be increased without a corresponding increase in the diameter of the fiber.

Each exceedingly limited section of neuronal membrane, whether squid or human, displays changes that are fascinating and rapid. A short section of nerve fiber is shown in Figure 11, where the inset indicates a few of the relevant details of the membrane. The membrane consists of a double layer of fatty (lipid) molecules that encase and enclose the neuron. Amid those lipid molecules, protein molecules serve to admit or refuse ions (through electrical-chemical-mechanical interactions that are very poorly understood) inside and outside the cell. As a battery draws ions from one chamber to another, the neuron continuously shuttles ions across the membrane. The reader should be aware that a real nerve axon is much more complicated than this simple sketch. Schematically, however, we can describe the manner in which the ions flow across the lipid-protein membrane as a marvel of electrical engineering. (Beyond that, it is also a subject of no small interest to anyone who wonders how the brain manages to *know* anything.)

The basic status quo (or resting state) is as follows: Potassium ions are found at relatively high concentrations inside the cell, where the electrical potential is negative, and relatively low levels outside the cell. Like salt water from an ocean seeping upriver, toward fresh water, the potassium ions naturally diffuse, pushing against the membrane toward the lower potassium concentrations outside the cell, but this diffusive tendency is balanced by the tendency of the potential difference across the membrane to pull potassium ions in, so the net flow of potassium ions is zero. At the same time, the sodium ion concentration is relatively low inside the cell and relatively high

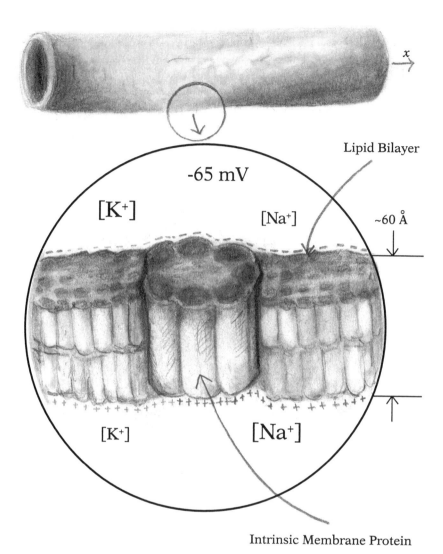

Figure 11. A short section of nerve axon at rest. The inset shows a protein that is imbedded in the nerve membrane. [K$^+$] *indicates the concentration of potassium ions, and* [Na$^+$] *indicates the concentration of sodium ions.*

outside it. Thus the sodium ions want to diffuse into the cell. The negative electrical potential inside (−65 millivolts) is also trying to pull them in. These two effects are not in balance (as for the potassium ions), and one is led to ask: Why don't sodium ions rush into the inside of the nerve? How can it remain at rest while the sodium ions are in this state of tension? The answer is that—at rest—the membrane permeability for sodium ions is very small, effectively zero.

We are next led to ask how the sodium ions get into their ostensibly "unhappy" state of high outside concentration, and what keeps them there. The standard (and glib) answer is that a *sodium pump*—driven by energy released in the breakdown of adenosine triphosphate, or ATP, the cell's version of gasoline—pushes these ions out against gradients of both ionic concentration and electrical potential. The sodium pump is almost certainly embodied in one of the aforementioned proteins lodged in the fatty membrane around the neuron, but the details of this are not at all understood at the present time.

We have so far described a nerve membrane at rest. But this begs a question: How does the spark ignite—how does the membrane generate a pulse of voltage?

Using axons of squids and other animals in the laboratory, and coupling such nerves to sensitive meters, electrophysiologists have made precise measurements of the ways that electric voltages and chemical concentrations change with time within neurons (Cole, 1968). It turns out that the switching action of the nerve membrane proceeds over a few thousandths of a second in the following manner. If—somehow—the voltage inside the membrane is changed from the resting level of -65 millivolts to about -40 millivolts, the membrane's permeability to sodium increases, and this increase permits sodium ions to begin to flow into the nerve. Since the sodium ions carry positive charge, their inward flow makes the inside voltage even less negative. (It is rather like starting a bonfire—a match lights some tinder, which begins to burn, releasing more heat that causes further burning—in a self-supporting dynamic process that electrical engineers like to call *positive feedback*.)

In nerves this ignition (or positive feedback process) continues for about a thousandth of a second (a *millisecond*) until the voltage inside the nerve becomes $+55$ millivolts with respect to the voltage outside. At this moment, which corresponds to the peak of the voltage curve in Figure 10, the inflow of sodium ions stops. Why does it stop? Not because the permeability of the membrane to sodium ions has fallen back to zero, for it has not. And not because the concentrations of sodium ions have changed, because they have not. Sodium ion current stops when the potential inside the membrane is $+55$ millivolts with respect to the potential outside because then the tendency of the concentration difference to push sodium ions inward is just balanced by the tendency of the voltage difference to push them outward.

But what of the potassium ions at this moment? And what happens to the membrane permeabilities when the inside is $+55$ millivolts with respect to the outside?

Over a time of a few milliseconds, two changes take place in the membrane. First, its sodium ion permeability falls back to zero, and second, the potassium ion permeability rises. Since the voltage inside the membrane is $+55$ millivolts, it tends to push potassium ions outward, and this now *adds* to the outward diffusion of potassium ions. Thus potassium ions carry positive charge out of the nerve, which brings the voltage inside back to its resting level of −65 millivolts. The sodium ions do not interfere with this return because their permeability has fallen back to zero.

From the perspective of an electrical engineer, this mechanism behaves like an electronic switch that jumps rapidly (on the order of milliseconds) from −65 millivolts to +55 millivolts and back again. Throughout this process, we can say, the total levels of potassium and sodium ions change little. In the language of engineering, the sodium and potassium concentration ratios act as stable batteries that drive the two phases of a monostable multivibrator. After repeated switchings of the membrane, the batteries begin to discharge (the ionic concentration ratios tend to decrease), but the original concentrations are restored by the sodium pump.

But what causes such a pulse of voltage to move along the axon?

NONLINEAR DIFFUSION

The process described above takes place at one location, one patch of membrane on the axon. As was noted, a metaphor for the process is the lighting of a bonfire or a watch fire, where all the fuel is located at one spot. Using an even simpler metaphor, it is like a single domino falling over. But suppose that domino is the first of a long row of dominos, carefully set up by a teen-aged enthusiast. Or suppose that the wood for the watch fire is spread out in a long line along a shore. Or perhaps someone has poured gasoline or gunpowder in a long strip and lit it at one end. As we all know, burning at one point heats the neighboring wood or gasoline or gunpowder, causing it to burst into flame, and the process of burning moves along the strip. It is this *process* of burning that is like a moving nerve *impulse*. The nerve conduction process drives itself forward, rolling along from the center of the neuron outward toward the tip of the axon. Then, as an impulse reaches the end of the axon, it arrives at a synapse—or a gap between neurons—and there sends a chemical signal to the dendrite of another neuron, triggering similar changes in an adjacent cell. And so the fire spreads, in many directions, throughout the brain.

Although there was much confusion about the nature of a nerve impulse in the nineteenth century, it is now clear that the diffusion of nerve conduction is not at all like traditional linear diffusion. Under the influence of linear diffusion, concentrations ebb and dissipate, much as a puff of smoke floats away on a still morning or the squid's inky pseudomorph vanishes into the sea. The nonlinear impulses that we have been examining release stored energy as they propagate, and this released energy supplies the power that is dissipated in the sputtering gunpowder or in the circulating ionic currents of the nerve impulse. Diffusion in the nerve is not linear at all. Nonlinear diffusion entails abrupt, wavelike impulses of activity that travel along the axon with constant speed and shape, carrying information reliably from one point to another (Fife, 1979).

All these changes have been studied in the laboratory in great detail. To cite the most famous instance, in 1938—shortly after the rediscovery that the giant axon of the squid is in fact a nerve—the American biophysicist Kenneth Cole used the newly invented cathode-ray oscilloscope to observe the changes with time of the voltage of a propagating nerve impulse (Cole and Curtis, 1938). This voltage appears as the solid line in Figure 10, while the width of the band (looking like a martini glass on its side) shows a measurement of the total membrane permeability, and the time marks indicate one-millisecond intervals. (The horizontal sweep rate of an oscilloscope was not so uniform in those early days.) Viewed as a function of time, the voltage rises to a peak, then subsides back to its resting level.

Let's try to connect this graph with the discussion of membrane switching. The sudden increase of membrane permeability marks the onset of sodium ion current—the moment that sodium ions, admitted by proteins in the neuronal membrane, rush into the cell. The foot of the voltage curve penetrates about 0.2 milliseconds ahead of the time when sodium current starts to flow (the top of the martini glass). This precursor changes the voltage inside the neuron from -65 millivolts to -40 millivolts, establishing the threshold condition that causes a new patch of membrane to switch. The peak of the voltage curve indicates that the inside of the nerve has reached a voltage of $+55$ millivolts with respect to the outside, so sodium current has stopped flowing. Finally, the slower relaxation from the peak voltage back to the resting voltage is dominated by an outward flow of potassium current.

THE HODGKIN-HUXLEY EQUATIONS

The influx of these ions and the very existence of the graph posed a fascinating challenge to biophysicists: Would it be possible to account for the behavior of the ions and the membrane mathematically with equations? Famously, the answer was yes.

In the early 1950s, Alan Hodgkin and Andrew Huxley made careful measurements of the sodium and potassium components of ionic currents of the giant axon of the squid under conditions in which the membrane voltage was constant both with respect to the space coordinate along the axon and with respect to time. In the jargon of electrophysiology, such an experiment is said to be *space-clamped* and *voltage-clamped*. Hodgkin and Huxley then constructed a set of phenomenological equations for the membrane current, which are presented and discussed in Appendix G. With these equations, Hodgkin and Huxley assumed the solution to the problem of ionic flow had the form of a traveling wave, which describes the process of continuous release of stored energy and its dissipation in the circulating currents that surround the nerve impulse. A traveling wave was chosen for their solution because it moves along with constant speed and unchanging shape, just as does a nerve impulse along a nerve. They were then able to compute (with a mechanical hand calculator!) a numerical solution to their equation that agreed with Figure 10, giving both the voltage curve (solid line) and the variation of the membrane permeability with time (the toppled martini glass). In addition, they correctly predicted the velocity of a squid nerve impulse. All of these predictions were obtained from their theory and space-clamped measurements of the properties of the membrane. Since their theory had no adjustable parameters, it had to be either right or wrong, and it was right (Hodgkin and Huxley, 1952).

NEURODYNAMICS

From thousands of experiments with isolated nerve fibers and the remarkable success of Hodgkin-Huxley theory, the nerve impulse emerges as a well-defined dynamic entity. On a particular fiber, it has a characteristic shape and speed. It is a *thing*, and asks to be treated as such. Some might find this assertion difficult to swallow, because a thing can also be a lump of mass or energy, like an electron or a molecule of benzene or a baseball or an electromagnetic pulse carrying the signal "dit" or "dah" from one ham radio operator's transmitter to another's receiver.

But, dear reader, do you not consider yourself to be a "thing," an object in the universe with an independent existence that would be taken seriously? Yet you are also a process that consumes and dissipates energy in the course of your existence. The atoms in your body change every few months, yet you—somehow—remain the same. So it is with the tornado that was described in the first chapter. So it is with a prairie fire. So it is with the flame of a candle and so with a nerve impulse.

Taking the nerve impulse seriously involves considering all aspects of its behavior. These include accounting for changes that arise when the diameter of an axon varies. If the thickness of a fiber gradually decreases in the direction that the impulse is moving, its velocity will decrease, while the opposite is true if the fiber diameter is gradually increasing. On the other hand, a *varicosity*, or bulge in the fiber, tends to cause a delay as the impulse passes through. Under appropriate conditions, this delay can become infinite, which is another way of saying that the passage of the impulse is blocked by the varicosity.

The Hodgkin-Huxley equations account correctly for all of these effects and more. They also forecast what happens when two nerve impulses collide in the lattice of neurons in the brain. When two nerve impulses experience a "head-on" collision (like engines on a poorly managed railroad), both are destroyed. Two impulses may also gently collide while going in the same direction, but in this case they become coupled together (like the freight cars of a well-managed railroad). This is what happens in the laboratory, and it is what happens in living tissue (Scott, 1977).

It is also interesting that in an experimental or physiological arrangement where two fibers are parallel and in close proximity, impulses can become transversely coupled into complex hyperpulses. This effect has been clearly observed experimentally and carefully studied theoretically, and it may be of importance in the functioning of bundles of nerve fibers such as the sciatic nerve, leading to the leg, or the corpus callosum, which links the two hemispheres of the brain (Scott and Luzader, 1979).

Finally, when the axons branch into two fibers, impulses may be delayed or blocked in a manner that is similar to the effects of varicosities noted above. Since a typical nerve cell has many branchings on both its output (axon) and input (dendrite) strands, such effects are important for the global functioning of a nerve cell and thus for the brain as a whole. The implications are discussed in greater detail in the next chapter.

Such evidence in hand indicates that the nerve impulse becomes a behavioral entity unto itself—an *atomic spark of thought*—and again we ask:

Can the Hodgkin-Huxley equations that govern nerve impulse dynamics be reduced to those of physics?

We know that they can be *traced* to the laws of physics because we have sketched the connections through chemistry, biochemistry, and the dynamics of membrane ion permeability. We know this despite passing over several barriers imposed by immense numbers of possibilities along the way. On the other hand, the equations are not "ordinary laws of physics" (as Schrödinger pointed out) but "new laws" that emerge at the hierarchical level of the axon to govern the dynamics of nerve impulses. One cannot derive these new laws from physics and chemistry because they depend upon the detailed organization of the intrinsic proteins that mediate sodium and potassium current across the membrane and upon the geometric structures of the nerve fibers.

We also see clearly, unmistakably, that the Hodgkin-Huxley equations serve as an example of biomathematical theory at its best. Despite repeated testing over the past 40 years, the integrity of the equations has never been seriously in doubt. The work has been established as the foundation of electrophysiology: it is the law that ordains the flow of ions across the membrane of the cell. As with Schrödinger's description of the hydrogen atom, there are *no* undetermined parameters to be adjusted by a desperate theorist. There is no need for animism, as Helmholtz had suspected, to explain the cerebral fires of a Romantic poet. Rather, the ions in your brain and my cat's brain flow across their neural membranes just as described by Hodgkin and Huxley.

Yet the atomic theory of Schrödinger and the Hodgkin-Huxley theory of nonlinear diffusion are quite different. Schrödinger's equation is based on the law of energy conservation, while the Hodgkin-Huxley equations don't conserve anything. Time, in Schrödinger's vision of the quantum theory, flows with equal ease in either direction, but solutions of the Hodgkin-Huxley system have a past, a present, and a future: they push forward in one direction with respect to time, just like our bodies and our thoughts. In other words, the dynamics of a nerve offer a striking illustration of the central premise of this book: that consciousness cannot be tightly and inextricably connected to lower levels of the scientific hierarchy. In the stairway to the mind, the atomic dynamics of the protein molecules in the neural membranes are unrelated to the electrical signals those proteins facilitate. This is not to diminish the quality of the science in either field, but only to recognize that one is not a logical outgrowth of the other.

From a more general perspective, both the quantum mechanical picture offered by Schrödinger and the powerful equations of Hodgkin and Huxley are among the most useful theories that science has to offer. Both have been subjected to hundreds of critical experimental tests—and passed. Yet they are not kin. There is no parameter in Appendix G (the Hodgkin-Huxley equations) that could be predicted or derived from the equations in Appendix B (Schrödinger's theory). None. To borrow from the language, if not the intent, of William James, "nerve stuff" is fundamentally different from "atom stuff."

So how can we consider a nerve impulse? Surprisingly enough, since the first nerve experiments were conducted by Galvani and Helmholtz, there has been an apt metaphor for this process: the candle. The candle explains more about what happens in nerve tissue than Schrödinger's marvelous equation.

Let us examine a few similarities. In accord with experiments, the Hodgkin-Huxley theory shows that a quantity of electric charge must be delivered to an axon in order to initiate a nerve impulse. Similarly, in order to light a candle, a certain temperature difference must be applied to the wick for a certain time. The same is true of nerve: without a trigger, the membrane will remain in its resting state. Moreover, once it has been lighted, the candle's flame advances along the wick in the same fashion that an impulse of energy pushes down the axon of the nerve. How do we know this? From the Hodgkin-Huxley equations.

The dynamics of a nerve impulse are like those of a flame, governed by the nonlinear diffusion equation. What's more, the nerve impulse and the candle flame share an all-or-nothing quality: they either burn or they do not, with obvious results in ordinary life. Finally, like the flame of a candle, the nerve impulse does not conserve energy. Instead it balances rates of energy release and loss. This idea is not so intuitive, but clearly the flame moves down the wick exactly as fast as the wax is consumed. Consider that the heat released by the flame diffuses into the melted wax and vaporizes it, supplying fuel to the flame which releases more heat, and so on. As the flame propagates along the candle—in a manner that is analogous to the propagation of a nerve impulse along the axon—a condition of power balance is established. The energy stored in the wax of the candle feeds energy to the flame at the same rate that it is consumed by the flame.

One of the earliest insights into the unique properties of nonlinear diffusion was expressed more than a century ago in a series of six lectures on the candle, which seem little-known to most who now investigate consciousness. The lecturer was Michael Faraday (1791–1867)—the British chemist and physicist responsible, among many other discoveries, for the principle of the electric dynamo. In these lectures, Faraday (1861) stated:

> There is no better, there is no more open door
> by which you can enter into the study of natural philosophy
> than by considering the physical phenomena of a candle.

It *is* amusing to think of scientists and philosophers of the nineteenth century sitting around a candlelit table pondering the nature of nerve. Yet Faraday's lectures, I believe, can today be soberly regarded as an uncanny prophecy for a century of investigation into the fundamental structure of the brain. It is impressively relevant more than a century after its utterance. Instead of finding an apparatus that is simply electrical, or merely chemical, electrophysiologists have indeed uncovered a biological wick, a thin fiber that flickers with extraordinary sensitivity. The human mind is built upon this natural wick, burning as it does in concert with millions of others. And as Faraday may have suspected, the neuron flickers on, then off, buffeted but never extinguished by the biological and chemical breezes to which it is exposed.

References

J M Arnold, W C Summers, D L Gilbert, R S Manalis, N W Daw and R J Lasek. *A guide to laboratory use of the squid* Loligo pealei. Marine Biological Laboratory, Woods Hole, 1974.

K S Cole. *Membranes, ions and impulses.* University of California Press, Berkeley, 1968.

K S Cole and H J Curtis. Electric impedance of nerve during activity. *Nature*, 142:209, 1938.

M Faraday. *A course of six lectures on the chemical history of a candle.* Harper, New York, 1861. (Republished in *The Harvard Classics*, P F Collier & Son, New York, 1910.)

P C Fife. *Mathematical aspects of reacting and diffusing systems.* Springer-Verlag, Berlin, 1979.

H Helmholtz. Messungen über den zeitlichen Verlauf der Zuchung animalischer Muskeln und die Fortpflanzungsgeschwindigkeit der Reizung in den Nerven. *Arch. Anat. Physiol.*, 276–364, 1850.

A L Hodgkin and A F Huxley. A quantitative description of membrane current and its application to conduction and excitation in nerve. *J. Physiol.*, 117:500–544, 1952.

A C Scott. *Neurophysics.* Wiley, New York, 1977.

A C Scott and S D Luzader. Coupled solitary waves in neurophysics. *Physica Scripta*, 20:395–401, 1979.

M W Shelley. Frankenstein *(or the Modern Prometheus).* Dell, New York, 1965 (first published in 1818).

Is There a Computer
In Your Head?

Membranes, webs of nerves that lay white and limp,
have filled and spread themselves
and float round us like filaments,
making the air tangible and catching in them
far-away sounds unheard before.

Virginia Woolf

I n an age when children use computers with nonchalance, when even the cheapest wristwatches contain rudimentary computers, it has almost become a cliche to reverse the popular comparison between brains and computers. We have heard for years that a watch has a "brain" on a silicon chip. But now, increasingly, we also read that the neurons of a human brain are themselves integrated circuits—and, to hear the boasts of some computer enthusiasts, not particularly fast ones at that! Even on the street, far from the university, one may hear that a brain is a machine, destined sooner or later to be overtaken by some piece of electronic equipment.

No profession has fostered the notion of a computer in your head more than the neuroscientists who study consciousness. It was the neuroscientists, after all, who documented the quasi-mechanical operation of a neuron. It was they who devised and verified the Hodgkin-Huxley equations that predict the behavior of a neuron. It was a sometime neuroscientist (myself) who compared the nerve membrane to an "electronic switch" in Chapter 4. It was even neuroscientists who noted similarities between the electrical current coursing through computers and the electrochemical signals rippling through the human brain.

But this neuroscientist, for one, is not convinced. Although there are similarities between computers and brains, there is—at this writing—no credible evidence to suggest that the ten billion neurons in the brain are equivalent to any computer we know. Indeed, even a single neuron is far more complex than many scientists in the field of consciousness will readily concede. Their assumptions, as we shall see here, are so simplistic as to be relatively useless in any discussion of how a real brain functions. *Any* man-made approximation of neurons in concert is roughly as complex as the knitting in a well-made sock: intricate, to be sure, but easily comprehended when held up for close inspection. Neurons in the brain, by comparison, are more properly regarded as the threads of a Persian carpet, thickly woven with connections between connections and structures within structures and variable patterns within variable patterns that elude understanding, much less duplication. How can that be so? How can the hype about computers be so inflated? To gain perspective, we must consider the McCulloch-Pitts neuron.

THE McCULLOCH-PITTS NEURON

In the previous chapter, we considered the axonal segment of the neuron. Here we shall consider the neuron in its entirety. Suppose that instead of aiding a squid in flight from a predator, the neuron in question has drawn the task of helping you know your watch is beeping. As simple as that beeping may seem, one neuron cannot deliver it all, process it all, digest it all. Many neurons must collaborate. But we can still approach that neuron (shown in Figure 12), from its dendrites to its axon and everything in between, as one component in the vast neural web allowing you to notice the beeping.

And what *is* the neurological chain of events that leads to your being conscious of its noise? In purely electrical terms, as we saw in the previous chapter, the response of one neuron is loosely similar to that of a transistor that regulates the temperature of your room by switching the furnace on when it is too cold and off when it is too hot. Input signals arrive at dendrites in the brain, and these signals somehow combine in the body of the neuron. The neuron may do nothing. Or it may send an impulse through to the axon, which in turn reaches dendrites of other neurons. If sufficient numbers of cells are activated, you notice the beeping. To the scientists who study such phenomena, the neuron "decides" whether to fire. If it does fire, the signal must pass down the axon, and naturally enough, the first part of the axon—the *initial segment* that emerges directly from the main core of the neuron—is of the greatest interest. If the electrical potential (voltage) on this segment of the axon rises above a threshold value, the axon fires. And if this firing occurs, the impulse travels toward the far twigs of the axonal tree, where contact is made to other neurons, generating other nerve impulses that pass through their dendrites, and eventually down more axons, and so on. The crux of the matter is not whether such signals occur; they do. The more fascinating problem is exactly how a single neuron regulates its dendrites and axons, deciding when to fire and when to hold the nerve impulses in abeyance. The mystery: What rules does a neuron obey?

Some of the insights are more than fifty years old. In the early 1940s, Warren McCulloch, then a professor of psychiatry and psychology at the University of Illinois Medical College, and Walter Pitts, a young mathematician, began to investigate individual neurons as they operated in concert with other neurons. Their approach was based upon the symbolic logic of Alfred North Whitehead and Bertrand Russell. In 1943 McCulloch and Pitts published a seminal paper on the analysis of networks composed of idealized neurons. This paper was all the more remarkable, in retrospect, because

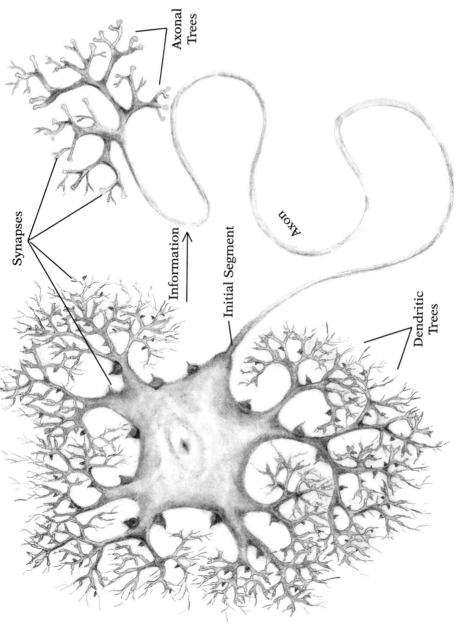

Figure 12. The structure of a typical neuron. Incoming information is gathered from synaptic contacts on the dendritic trees. After processing, this information is carried by nerve impulses to the tips of the axonal trees.

at the time there were no digital computers, there was no electronic logic, there were no mainframes or workstations or silicon chips to stimulate the imagination of a neuroscientist.

Adhering to the counsel of William of Ockham, whose metaphorical razor is mentioned in the introduction to this book, their conjectural neurons were cut to the essence. McCulloch and Pitts assumed the neurons to have just three properties. First, the activity of the neuron would be an "all-or-none" process. It either fires or it doesn't; there is no state in between. Second, a certain fixed number of input signals were assumed to arrive within a certain time period (the period of latent addition) for the model neuron to fire. The time delay for propagation of a signal through the neuron, finally, was assumed to have a constant value. That assumption ensured a consistency in the computations of the signals that would be difficult to obtain otherwise.

The McCulloch-Pitts neuron, as it turns out, has become the workhorse of neural network theory, in which scientists attempt to construct computers composed of individual components interacting among themselves. Fifty years later, scientists in the field often make the *same* assumptions about neurons, even when they are—or should be—aware of electrophysiological research that shows those assumptions to be greatly oversimplified. The problem is not with McCulloch and Pitts; they were pioneers. It is with the myriad latter-day scientists who are not sufficiently concerned about the subtleties within a single neuron, and the degree to which those subtleties make a McCulloch-Pitts neuron appear almost as a caricature of an actual neuron.

As envisioned by McCulloch and Pitts, their all-purpose neural element computes a linear combination of the input signals and compares it with a threshold level at the initial segment of the axon. If the linear sum is below the threshold of that particular cell, the output remains at rest. The neuron does nothing. But if it exceeds threshold, the axon fires, and a nerve impulse propagates toward the tips of the axonal tree. The only nonlinear aspect of the model is the comparatively simple decision to fire. A threshold firing decision *is* nonlinear, the reader will recall, because individual inputs acting alone do not cause an output signal, but acting together they do.

As much as scientists would like to arrive at a reductionist perspective of the neuron, isolating one crucial component that forms the decision to fire becomes more and more difficult as the years go by and experimental evidence from laboratories of electrophysiology accumulates. The neuron is more than a single switch. It seems to be composed of many smaller switches located throughout both its dendritic and its axonal branches.

DENDRITIC AND AXONAL COMPUTING

First let us consider the dendrites, the in-basket filaments which, according to classical neurophysiology, passively gather signals from adjacent axons or sense organs and funnel them to the main body of the neuron. This, of course, was the assumption of McCulloch and Pitts, who could not have known otherwise, and of many theorists in our own day, who should know better. The reason that early electrophysiologists assumed all dendrites to be passive was because they were overly impressed by the all-or-none behavior of the active nerve impulse, which was described in the previous chapter. If dendritic impulses were not passive but active—so the argument ran—then each incoming impulse at the tip of a dendritic tree would be able to ignite the entire tree, and it would not be possible for the dendrites to sum their inputs.

As one of the more impressive examples of dendritic arborization, let us look at the Purkinje cell of the human cerebellum (shown in Figure 13), which was studied by the Spanish histologist Ramón y Cajal (1952). Espaliered over an area of about a tenth of a square millimeter—like a fruit tree against the warm southern wall of a Mediterranean villa—these dendrites receive not sunshine but the inputs from about 80,000 other neurons. The output signal, which travels along the axon, guides the muscles of an athlete or an artist. But why, one must ask, are there so many branches in these dendritic trees?

The picture that emerges from a careful analysis based on the Hodgkin-Huxley equations is rather different from the simple, classical picture. This description shows that active, all-or-none impulses do not spread, like a forest fire, out of control through the dendritic trees. They can be *blocked* where the dendrites branch (or bifurcate) as shown in Figure 14. An active impulse on one of the daughter branches does not automatically propagate through to the parent; instead, the dendrites themselves can make many independent decisions to fire—just as the classical electrophysiologists assumed was the case in a neuron as a whole (Berkinblit, et al., 1971; Llinás and Nicholson, 1971; Llinás et al., 1969). The reader should be aware that this is not an odd or isolated assertion. Based on established properties of the Hodgkin-Huxley equations, it has been verified by many scientists from several theoretical perspectives (Arshavskii et al., 1965; Bogoslovskaya et al., 1973; Gutman and Shimoliunas, 1973; Guttman, 1971; Patushenko et al., 1969; Scott, 1973a,b).

With reference to Figure 14, the basic idea is to consider the conditions for establishing an impulse on the parent fiber (C) as a result of incoming impulses on the daughter fibers, A, B, or both. Whether an impulse forms on C depends upon whether the threshold conditions for impulse formation on C are satisfied, and this in turn depends upon the geometry of the branch: the ratios of the sizes of the daughter branches to that of the parent. If these ratios are such that an incoming impulse on either daughter is able to fire the parent, the branch acts as an OR gate. In other words, A OR B can fire C. If incoming impulses on *both* daughters are required, the branch acts as an AND gate: both A AND B are required to fire C. In this way, a single dendritic branch is able to compute (or perform elements of Boolean logic) as is described in Appendix H.

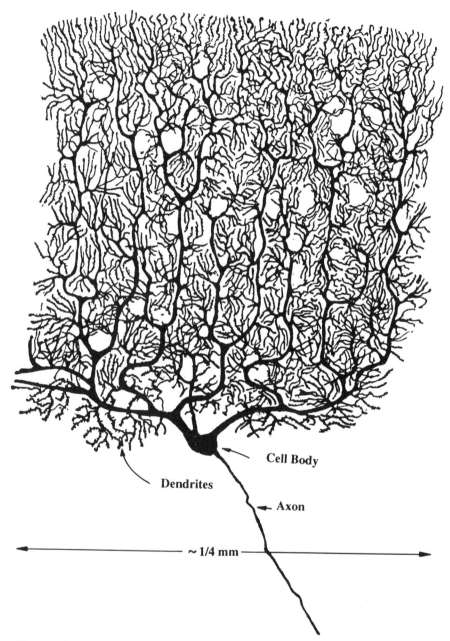

Figure 13. *A Purkinje cell of the human cerebellum. This cell receives about 80,000 synaptic inputs. (After Ramón y Cajal, 1952.)*

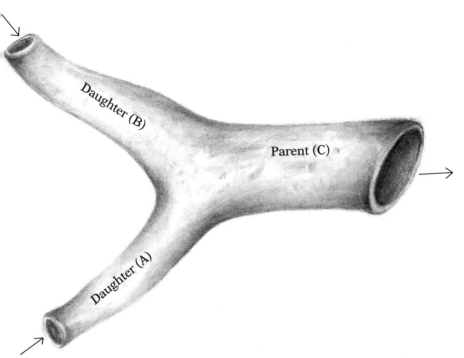

Figure 14. *Geometry of a branching dendrite. Can incoming impulses on either or both of the daughter branches ignite an impulse on the parent?*

Direct evidence for such computations appeared in 1963, with observations of impulse generation in branching regions of central neurons in mollusks (Tauc and Hughes, 1963). Circumstantial evidence for this point of view is provided by observations of intermittent bulges (or *varicosities*) in the dendrites of auditory (cochlear) neurons in various vertebrate species (Bogoslovskaya et al., 1973), as is shown in Figure 15. We learned in the previous chapter that such a varicosity introduces a time delay into the propagation of a nerve impulse. Upon sensing a beeping noise, the cochlear neurons seem to interject delays in the transmission of impulses. This suggests that the dendrites themselves, and not only the center of the neuron, play a significant role in the processing of auditory information.

On the axonal (or out-basket) side of the neuron, there has long been evidence that the classical picture of a single impulse on the main trunk spreading to every tip of the axonal tree might be overly simplified. As early as 1935, Barron and Matthews observed conduction block on the spinal cord of the cat, and in 1959 Krnjević and Miledi described failures of pulse transmission in motor axons of the rat. Were these early reports mere oddities, or did they suggest something important about the function of the axonal tree?

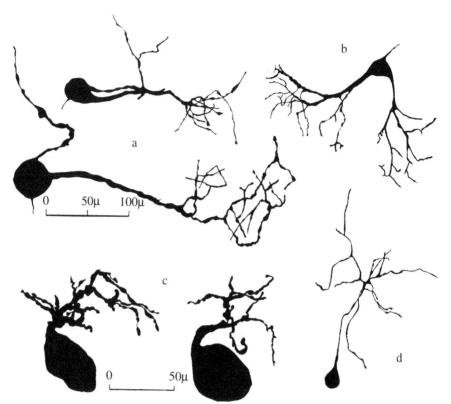

Figure 15. *Cochlear neurons of (a) monkey, (b) hedgehog, (c) owl, and (d) bat. (From Bogoslovskaya et al., 1973.)*

It fell to a brilliant electrophysiologist at MIT to provide an answer to these questions. Jerome Lettvin has spent many years studying information processing on the optic nerves of various species, often with great success. Together with McCulloch and Pitts, he published in 1959 a classic paper entitled "What the frog's eye tells the frog's brain," and a decade later he was involved in similar experiments on the optic nerve of the cat. One would presume, according to the classic McCulloch-Pitts view of neurons, that the acuity of the cat's vision can be traced, in part, to the fact that information gathered in its visual field reaches the brain completely and accurately. Put another way, it would seem logical that cells in the eye would fire reliably into certain cells in the brain, and that those cells would fire into other cells, and so on. But that is not what Lettvin and his colleagues learned (Chung et al., 1970). In a departure from the conventional thinking in neurobiology, his experiments show that signals from the cat's retina do not automatically proceed to the tip of every branch of the axonal tree of the optic nerve. Impulses generated in the eye do not necessarily even *reach* the brain. The cat's optic nerve has a means of regulating which of a particular set of impulses will go to which twigs of its tree.

Thus the axon is able to analyze the time between impulses, or perhaps their sizes, before sending the information onward. Lettvin's conclusion from these observations was striking. *A real axon has the ability to translate a time code on the trunk of the axonal tree (embodied in the temporal intervals between impulses) into a space code at the twigs of the tree (embodied in the particular twigs that are excited).*

This proposal for "multiple meanings" on single axons of the optic nerve was met with skepticism—and even some hostility—because it greatly complicates the task of developing a science of the mind, but confirming reports soon began to appear in the journals. Similar blockages and steering of information were observed on axons in the abdomen of the crayfish (Parnas, 1972) and the lobster (Grossman et al., 1973, 1979a,b), on sensory neurons of the leech (Van Essen, 1973), and in the walking leg of the crayfish (Hatt and Smith, 1975; Smith and Hatt, 1976). Also, changes in conduction velocity and excitability were observed on visual callosal axons of the rabbit following stimulation by multiple impulses (Swadlow and Waxman, 1976).

Here again, the reality of the neuron in action is richer and more varied than the standard assumptions posit. In studies of real neurons, the branches of axons emerge as regions of low *safety factor* where it is relatively easy for a nerve impulse to be extinguished. If we adopt the candle metaphor of the previous chapter, the flame of the neuronal impulse is quietly snuffed out. Here again, the departure from the simplistic classical view of the neuron as a single switch is rather stark: the old picture is that the axon will always transmit the impulse; the new one is that it sometimes does so, but on other occasions, independent of the core of the neuron, the axon by itself "elects" not to fire. Why does axonal blocking occur? Branching conductance may be influenced by changes in local geometry (Smith, 1980b), but an important factor is often an accumulation of external potassium ions, which leads to a depolarization of the membrane and reduced safety factor (Smith, 1977, 1980a). This is an effect that can be calculated from the Hodgkin-Huxley equations (Adelman and Fitzhugh, 1975).

Impressed by such evidence for information processing at the branching regions of dendritic and axonal trees, and convinced of its importance for the functioning of a brain, Uja Vota-Pinardi, an Italian biologist, and I organized and conducted a year-long series of experiments on the branching regions of squid axons. The broad aims of these experiments were twofold. First of all, we wanted to understand how nerve impulses propagate when they are crowded closely together, how their speeds and safety factors decrease. Secondly, we wished to study how pairs of closely spaced impulses would propagate through branching regions. (In candor, I should admit that the prospects of spending a year at the *Stazione Zoologica* on the Bay of Naples and sharing the atmosphere and *cucina* of that fascinating city were contributing factors.)

Working with the squid *Loligo vulgaris*, sacrificing two beautiful animals each morning and often continuing far into the night, we soon observed a decrease of 20% to 30% in the velocity for impulse spacings of several milliseconds (Scott and Vota-Pinardi, 1982a). Such a strong decrease in the impulse speed indicates a greatly reduced safety factor (or margin of stability against extinction). We then passed on to studies of the propagation of pairs of impulses through branching regions, as is shown in Figure 16 (Scott and Vota-Pinardi, 1982b). In this figure, the point S is stimulated. The resulting nerve impulse is recorded above the branch at electrode B and below the branch at electrode A. Glass microelectrodes (filled with ionic fluid) were used to measure the voltage passing each point, and create the figures shown here.

A comparison of Figure 16 with Figure 10 shows that standard Hodgkin-Huxley impulses were readily observed. In Figure 17, however, the preparation is stimulated by two impulses that are slightly closer together in the top recording, where the second impulse is blocked at the branch, than in the bottom one, where it is able to pass through. It is interesting to notice, in the upper trace of the bottom image in Figure 17, that firing at the branch introduces a distinctive hump into the observation at electrode B. There is no doubt that the hump, seen at electrode B, is related to the passage of the second impulse, which is observed on electrode A. One can carefully adjust the impulse spacing to repeatedly go back and forth between the upper and lower images of Figure 17. But *why* is this extra hump seen at electrode B?

When the branch decides to switch, as it turns out, evidence of that switching travels back along the daughter to electrode B. It is the difference between the time delay for this effect and the time delay for the second signal to travel from the branch to electrode A that is observed as the interval called T_D in Figure 17. Thus electrode B shows both the second impulse on its way toward the branch and the firing of the branch to form the second impulse on the parent. It took me several days to sort this out.

In the course of these experimental studies, I was often struck by how challenging it is to understand the dynamics of only two impulses that are traveling past a single branch of a fiber. How much more difficult, one wonders, must it be to comprehend the dynamics of an entire neuron? Or a brain?

SYNAPTIC TRANSMISSION

There is also rich and yet baffling activity in the joints between neurons: the synapses. At the twigs of its axonal tree, one may recall, every nerve cell abuts muscle cells or other neurons over a minuscule gap. Figure 18 shows a typical neuronal canyon (called the *synaptic cleft*) between an axon of one neuron and the dendrite of another. Within the end of the bulb (or *bouton*) at one side of the synapse are several hundred synaptic vesicles (about 500 Å in diameter), each of which contains several thousand transmitter molecules that can either depolarize (excite) or hyperpolarize (inhibit) the postsynaptic membrane, inducing or blocking nerve impulses on the other side of the gap.

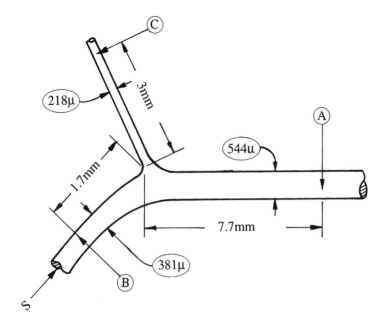

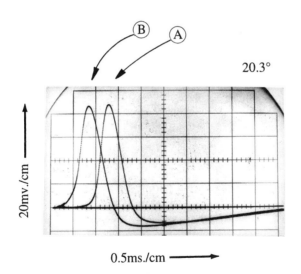

Figure 16. *Branching geometry (upper) and single action potentials (lower) for propagation on a squid giant axon. (From Scott and Vota-Pinardi, 1982b.)*

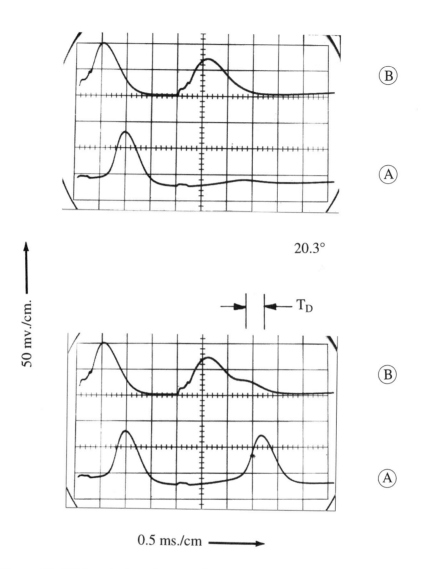

Figure 17. With twin-impulse stimulation at (S), the second impulse fails (upper) and passes (lower). (From Scott and Vota-Pinardi, 1982b.)

Upon arrival at a synapse, a nerve impulse causes the bouton at the very end of the axon to become flooded with calcium ions. The calcium ions, in turn, cause the vesicles to release transmitter molecules into the synaptic cleft, a process called *exocytosis*. These transmitter molecules then float across the cleft to the postsynaptic membrane via the process of linear diffusion, as discussed in Appendix F, with a time delay of about 300 milliseconds (Eccles, 1964).

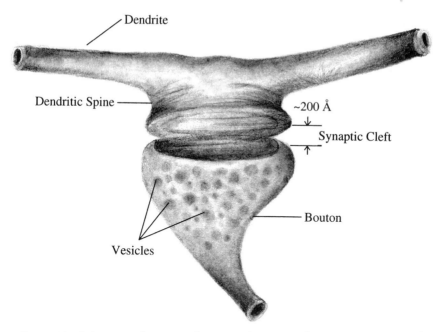

Dendrite

Dendritic Spine

~200 Å

Synaptic Cleft

Bouton

Vesicles

Figure 18. Schematic diagram of a synaptic contact between an axon and a dendritic spine. The vesicles are floating inside the bouton.

THE MULTIPLEX NEURON

Roughly twenty years ago, the new view of the neuron assumed some currency. We now may expect a nerve cell to be at least as complex as the *multiplex neuron* suggested by neurologist Stephen Waxman (1972) and sketched in Figure 19. He described four different regions of information processing in a single cell:

1. The *dendritic region* where both excitatory and inhibitory synaptic inputs are summed and logical decisions are made at the branches, as indicated by the question marks.
2. The *nerve body* and the *initial axon segment*, which has a low threshold for firing. Even the initial segment may receive synaptic input to assist in its decision to fire the axon.
3. The *axonal tree*, which is often covered by a myelin sheath that restricts membrane current to active nodes and thereby speeds conduction. These nodes can receive synaptic inputs, and information processing may occur at branches.
4. The synaptic outputs, which can be modified by input contacts from other cells.

In the terms of computer engineering, McCulloch and Pitts had viewed the individual neuron as a "gate" composed of a single transistor. From the

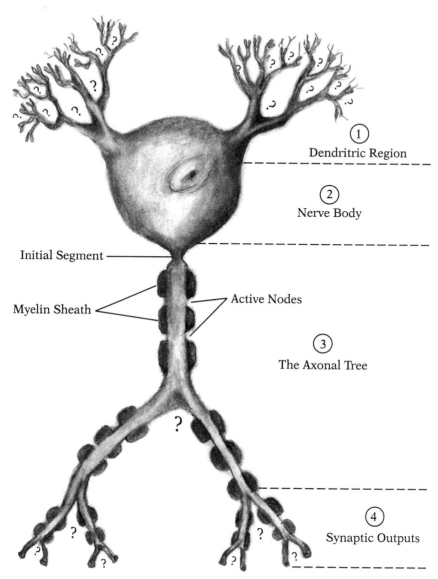

Figure 19. *A cartoon of the multiplex neuron. Question marks indicate branching regions of low safety factor that may perform logical operations. The myelin sheath speeds axonal conduction. Note that this figure is not to scale. (After Waxman, 1972.)*

perspective of Waxman's multiplex neuron, however, it looks more like a "chip" or an integrated circuit that is composed of many interconnected transistors.

A REAL NEURON

Although a striking and original concept at the time of its introduction, Waxman's multiplex neuron is now, in scientific terms, in early middle age. The weight of accumulating experimental evidence not only supports the multiplex neuron concept but suggests that a real neuron may be even more complex.

In 1976, for example, Schmitt, Dev, and Smith noted that our current state of knowledge concerning neuronal circuitry is undergoing a "quiet revolution." In their words:

> The new view of the neuron, based primarily on recent electron microscope evidence and supported by intracellular electrical recording, holds that the dendrite, far from being only a passive receptor surface, may also be presynaptic, transmitting information to other neurons through dendrodendritic synapses [synaptic connections between dendritic branches]. Such neurons may simultaneously be the site of many electrotonic current pathways [direct electrical connections], involving components as small as dendritic membrane patches or individual dendrites.

These dendrodendritic synapses may be sites for interactions between the cytoskeleton and the membrane potential.

Moreover, in a term suggested by Llinás (1988), the neuron should no longer be viewed as a "Platonic" device in which dendritic inputs are chastely funneled to the initial segment via passive pathways. Rather, as Adams (1992) has recently put it, the Platonic neuron has "got the hots" from a wide variety of experiments showing the effects of both calcium ion and sodium ion channels in dendrites. In short, these filaments act as nonlinear amplifiers to boost distal signals in specific ranges of voltage and frequency, in addition to supporting full-blown action potentials. Thus the dendrite cannot be considered a mere passive mechanism to produce a linear sum of the input signals. Similar effects have recently been observed and analyzed in the photoreceptors of drone honey bees (Vallet et al., 1992).

Finally, I note recent evidence that the synapse itself is multifunctional, not only influencing firing decisions over a latent time period but actually changing the information processing schedule in a neuron. In the dentate granule cells of the hippocampus, for example, the activity of a group of synapses can depress the strengths and capacity for plasticity of their neighbors for periods up to a quarter hour (Stevens, 1993). It is not yet clear how to incorporate such effects into neural models.

In summary, although electrophysiologists perform experiments in which various stimuli are introduced as dendritic inputs and recorded as axonal outputs, Jerry Lettvin has pointed out that we don't know the "language of the brain" so we can't even distinguish between signal and noise (Chung et al., 1970). This should be recognized as a serious problem by those who attempt to make computer models of real neurons. Instead of a normal pasture gate that is opened and closed by a lad or lass, we must now imagine one that is surrounded by an unruly mob of villagers involved in an eternal argument over whether to open it or keep it closed.

When a philosopher or a computer scientist makes statements about the brain that are based upon a model of McCulloch-Pitts neurons, one must be concerned about oversimplification. What claims can be made about the dynamics of a system that is composed of ten billion minicomputers?

QUANTUM EFFECTS?

To this point, we have been discussing neuron dynamics entirely in the context of classical theory and its causal dynamic laws. Quantum effects, which dominate the dynamics of an individual atom, have been ignored. This is consistent with the prevailing views of most electrophysiologists, and it is a reasonable assumption from the perspective of Heisenberg's uncertainty principle. This principle, the reader will recall from Chapter 2, states that the product of uncertainties of classical measurements of energy and time can be no less than Max Planck's famous constant (6.626×10^{-34} joule-seconds).

In a neural experiment (see Figures 16 and 17), time intervals are of the order of 10^{-4} seconds and the threshold energy to fire a pulse on a squid axon is about 10^{-11} joules (Scott, 1973a), so the product of the energy and the time is 10^{-15} joule-seconds. Since 10^{-15} is larger than Planck's constant by a factor of more than 10^{18}, the error or uncertainty in classical descriptions of time and energy could be as small as one part in $\sqrt{10^{18}} = 10^9$. In other words, both energy and time could be simultaneously measured to an accuracy of about one part in a billion (10^9) without violating the quantum mechanical uncertainty principle. An accuracy of one part in a billion is so far beyond anything achieved in the electrophysiologist's laboratory that it would seem quantum corrections to nerve measurements are not necessary.

Nonetheless, it has recently been suggested that quantum effects *do* play a role in the dynamics of a nerve cell. These suggestions are not inspired by experimental observations but by a seeming contradiction between the observed nature of human consciousness and the logical constraints of reductive science. Since the resolution of this conflict is an aim of the present book, it is appropriate to consider the proposed mechanisms for quantum influence on neural behavior.

Henry Stapp (1993) argues that nothing in classical physics is more than the sum of its parts. He insists that classical physics is logically complete, and therefore it has no place for such mental phenomena as consciousness. But, as we all know from ordinary life, consciousness does exist; thus classical physics may be inadequate to describe the brain. Inspired by Stapp's work, John Eccles recruited the assistance of Friedrich Beck, a quantum physicist, to develop a quantum theory of exocytosis in the synapse. According to this theory (Beck and Eccles, 1992), the probability of vesicle emission into the synaptic cleft is influenced by some quantum process—perhaps the movement of a hydrogen bond. In a similar vein, Roger Penrose, a theoretical physicist, and Stuart Hameroff, a practicing physician with a strong bent for basic research, have proposed that quantum effects in the cytoskeleton may play a role in the global activity of a neuron (Hameroff, 1994; Penrose and Hameroff, 1995). As with everything in science, such theories must ultimately stand or fall on the basis of experimental evidence.

CAN A NEURON BE MODELED ON A COMPUTER?

If we assume that the dynamics of a neuron are deterministic (output signals can be predicted from the input signals), then the answer is "yes," but this may be a trivial answer. The computer representation of a particular neuron would certainly be very complicated, and we should not forget that each neuron is different, each is an individual. Thus it might be impractical to construct a model of all the neurons in a realistic nerve system. Whether or not this is so depends, of course, on the level of technology that is available. Much more powerful computers can be built now than in 1943, when the computer was being invented, and corresponding developments are to be expected in the future, especially as one considers the growing interest in computing devices that are constructed at the molecular level.

On the other hand, the dynamics of a neuron might be influenced by quantum effects, either in the synapse or the microtubular interior, implying that output signals could be related to input signals only in a probabilistic manner. In this case the answer to the above question would be "no" because a quantum system is not deterministic. Only the most ardent and optimistic computer engineers are contemplating machines in which quantum effects play a decisive rôle.

The humble neuron, in a sense, emerges with what almost appears to be a mind of its own. Sometimes, for reasons not yet clear, it fires predictably, dependably. But on other occasions, quixotically, it pauses or elects not to fire, sitting quietly, marshaling chemical and electrical resources for release later. All this is not to say that up to a point the classical interpretations of the neuron are wrong; they remain useful. But no serious student of consciousness should bet the family farm that a solitary neuron is a purely mechanical entity.

References

P R Adams. The Platonic neuron gets the hots. *Curr. Biol.*, 2:625–627, 1992.
W J Adelman and R Fitzhugh. Solutions of the Hodgkin-Huxley equations modified for potassium accumulation in a periaxonal space. *Fed. Proc.*, 34:1322–1329, 1975.

Yu I Arshavskii, M B Berkinblit, S A Kovalev, V V Smolyaninov, and L M Chailakhyan. The role of dendrites in the functioning of nerve cells. *Dokl. Akad. Nauk SSSR*, 163:994–997, 1965.

D H Barron and B H C Matthews. Intermittent conduction in the spinal cord. *J. Physiol.*, 85:73–103, 1935.

F Beck and J C Eccles. Quantum aspects of brain activity and the role of consciousness. *Proc. Natl. Acad. Sci. (USA)*, 89:11357–11361,1992.

M B Berkinblit, N D Vvedenskaya, L S Gnedenko, S A Kovalev, A V Kholopov, S V Fomin, and L M Chailakhyan. Interaction of nerve impulses in a node of branching (Investigations of the Hodgkin-Huxley model). *Biophysics*, 16:105–113, 1971.

L S Bogoslovskaya, I A Lyubinskii, N V Pozin, Ye V Putsillo, L A Shmelev, and T M Shura-Bura. Spread of excitation along a fiber with local inhomogeneities. *Biophysics*, 18:944–948, 1973.

S H Chung, S A Raymond and J Y Lettvin. Multiple meaning in single visual units. *Brain. Behav. Evol.*, 3:72–101, 1970.

J C Eccles. *The physiology of synapses*. Springer-Verlag, Berlin, 1964.

D C Van Essen. The contribution of membrane hyperpolarization to adaptation and conduction block in sensory neurons of the leech. *J. Physiol.*, 230:509–534, 1973.

Y Grossman, I Parnas, and M E Spira. Differential conduction block in branches of a bifurcating axon. *J. Physiol.*, 295:283–305, 1979a.

Y Grossman, I Parnas, and M E Spira. Ionic mechanisms involved in differential conduction of action potentials at high frequency in a branching axon. *J. Physiol.*, 295:307–322, 1979b.

Y Grossman, M E Spira, and I Parnas. Differential flow of information into branches of a single axon. *Brain Res.*, 64:379–386, 1973.

A Gutman and A Shimoliunas. Finite dendrite with an N-shaped current-voltage characteristic for the membrane. *Biophysics*, 18:1013–1016, 1973.

A M Guttman. Further remarks on the effectiveness of the dendritic synapses. *Biophysics*, 16:131–138, 1971.

S R Hameroff. Quantum consciousness in microtubules: An intra-neuronal substrate for emergent consciousness? *J. of Consciousness Stud.*, 1:91–118, 1994.

H Hatt and D O Smith. Axon conduction block: Differential channeling of nerve impulses in the crayfish. *Brain Res.*, 87:85–88, 1975.

K Krnjević and R Miledi. Presynaptic failure of neuromuscular propagation in rats. *J. Physiol.*, 149:1–22, 1959.

J Y Lettvin, H R Maturana, W S McCulloch, and W H Pitts. What the frog's eye tells the frog's brain. *Proc. IRE*, 47:1940–1959, 1959.

R Llinás. The intrinsic electrophysiological properties of mammalian neurons: Insights into central nervous function. *Science*, 242:1654–1664,1988.

R Llinás and C Nicholson. Electrophysiological properties of dendrites and somata in alligator Purkinje cells. *J. Neurophysiol.*, 34:532–551, 1971.

R Llinás, C Nicholson, and W Precht. Preferred centripetal conduction of dendritic spikes in alligator Purkinje cells. *Science*, 163:184–187, 1969.

W S McCulloch and W H Pitts. A logical calculus of the ideas immanent in nervous activity. *Bull. Math. Biophys.*, 5:115–133, 1943.

I Parnas. Differential block at high frequencies of branches of a single axon innervating two muscles. *J. Neurophysiol.*, 35:903–914, 1972.

V F Pastushenko, V S Markin, and Yu A Chizmadzhev. Branching as a summator of nerve pulses. *Biophysics*, 14:1130–1138, 1969.

R Penrose and S R Hameroff. Quantum computing in microtubules: Self-Collapse as a possible mechanism for consciousness. In *Toward a science of consciousness*, S R Hammeroff, A W Kaszniak, and A C Scott, editors. MIT Press, Cambridge, Massachusetts, 1995.

S. Ramón y Cajal. *Histologie du système nerveux*. Cons. Cup de Invest. Scientíficas, Madrid, 1952.

A C Scott. Strength duration curves for threshold excitation of nerves. *Math. Biosci.*, 18:137–152, 1973a.

A C Scott. Information processing in dendritic trees. *Math. Biosci.*, 18:153–160, 1973b.

A C Scott and U Vota-Pinardi. Velocity variations on unmyelinated axons. *J. Theoret. Neurobiol.*, 1:150–172, 1982a.

A C Scott and U Vota-Pinardi. Pulse code transformations on axonal trees. *J. Theoret. Neurobiol.*, 1:173–195, 1982b.

F O Schmitt, P Dev, and B H Smith. Electrotonic processing of information by brain cells. *Science*, 193:114–120, 1976.

D O Smith. Ultrastructural basis of impulse failure in a nonbranching axon. *J. Comp. Neur.*, 176:659–670, 1977.

D O Smith. Mechanisms of action potential propagation failure at sites of axon branching in the crayfish. *J. Physiol.*, 301:243–259, 1980a.

D O Smith. Morphological aspects of the safety factor for action potential propagation at axon branch points in the crayfish. *J. Physiol.*, 301:261–269, 1980b.

D O Smith and H Hatt. Axon conduction block in a region of dense connective tissue in crayfish. *J. Neurophysiol.*, 39:794–801, 1976.

H P Stapp. *Mind, matter, and quantum mechanics*. Springer-Verlag, Berlin, 1993.

C F Stevens. Reworking an old brain. *Curr. Biol.*, 3:551–553, 1993.

H A Swadlow and S G Waxman. Observations on impulse conductions along central axons. *Proc. Natl. Acad. Sci. USA*, 72:5156–5159, 1975.

H A Swadlow and S G Waxman. Variations in conduction velocity and excitability following single and multiple impulses of visual callosal axons in the rabbit. *Exp. Neurol.*, 53:128–150, 1976.

L Tauc and G M Hughes. Modes of initiation and propagation of spikes in the branching axon of molluscan central neurons. *J. Gen. Physiol.*, 46:533–549, 1963.

A M Vallet, J A Coles, J C Eilbeck, and A C Scott. Membrane conductances involved in amplification of small signals by sodium channels in photoreceptors of drone honeybee. *J. Physiol.*, 456:303–324, 1992.

S G Waxman. Regional differentiation of the axon: A review with special reference to the concept of the multiplex neuron. *Brain Res.*, 47:269–288, 1972.

V Woolf. *The Waves*. Harcourt, Brace, Jovanovich, 1978.

"An Enchanted Loom"

i cry no quarter of my age and call
on coming wits to prove the truth
of my stark venture into fate's cold hall
where thoughts at hazard cast the die for sooth

Warren S. McCulloch

I n Chapter 3, the subject of biochemistry was introduced through a metaphor that has always fascinated me: fireflies communicating with each other on a summer evening by switching on and off their mysterious lanterns of luciferase. A similar metaphor was used by Charles Sherrington—the leading neurophysiologist of his time—to describe the brain's behavior. In his presentation of the 1937–38 Gifford Lectures at the University of Edinburgh, the waking of the brain was pictured as follows:

> The great topmost sheet of the mass, that where hardly a light had twinkled or moved, becomes now a sparkling field of rhythmic flashing points with trains of travelling sparks hurrying hither and thither. The brain is waking and with it the mind is returning. It is as if the Milky Way entered upon some cosmic dance. Swiftly the head mass becomes an enchanted loom where millions of flashing shuttles weave a dissolving pattern, always a meaningful pattern though never an abiding one; a shifting harmony of subpatterns.

In 1940 these lectures were published as a book entitled *Man on his nature*, in which the knowledge and wisdom of an eminent neuroscientist are enriched by the style of a poet to trace the development of mankind's self-knowledge throughout history (Sherrington, 1951).

Thus the 1940s began. During this turbulent decade—amid the suffering and waste of war and the collapse of empires—events of interest to our little story also transpired. The first digital computers were constructed, and solid state electronics entered a phase of active development leading to the invention of the transistor in 1950. For students of the brain, two publications appeared that were of fundamental importance. The first of these, by Warren McCulloch and Walter Pitts, we have already met: a paper entitled "A logical calculus of the ideas immanent in nervous activity" (1943). The

75

second, in 1949, was Donald Hebb's classic book *The organization of behavior*. Neither of these seminal works was influenced by the other, yet both were indebted to Charles Sherrington. Both asked the question: How are the observations of psychology related to the dynamics of the brain's neurons? To understand their answers to this question, we must recall some of what had been learned about the structure of the brain since the middle of the nineteenth century.

A BRIEF LOOK AT THE BRAIN

Like a country in which certain regions harvest different crops, the brain is partitioned by task. In general, the scientific discovery of what portion of the brain is responsible for which function has advanced by comparisons between an injury to a certain part of the brain and a resulting change in behavior. The most famous example was identified, in 1861, by the French anthropologist and surgeon Paul Broca, who treated battlefield survivors and discovered that a small section of tissue on the side of the head controlled our ability to speak.

Broca's discovery would provide glimmers of insight into a larger truth about the brain: that the most obvious facet of its organization is into two equal hemispheres. The left half of the brain controls the right side of the body, and the right side of the brain controls the left side of the body. In most individuals, complex tasks such as counting and speech reside in one half of the brain. Thus in right-handed people (for whom the left side of the brain gives orders to the dominant hand), a stroke on the left side of the head will result in the loss of speech. A stroke on the right side of the head has other effects.

Beyond the left-right division in the brain, there are smaller landmarks of equal significance. Immediately above the spinal cord, for example, lies the most primitive part of the brain, the brain stem and cerebellum, tissues most similar to those in other vertebrates. Automatic functions such as breathing, heartbeat and digestion are governed here. In the middle of the organ, nestled at the center of the head, are glands governing growth, hormone levels, and reproduction. Finally comes the third major section of the brain, the wrinkled outer rind or *neocortex*.

Even to the novice observer of the neocortex, deep creases in the gray outer layer of each hemisphere clearly delineate four additional regions, known as the frontal, temporal, parietal, and occipital (or optic) lobes. The occipital lobes, at the rearmost tip of each hemisphere, control vision, for they process signals from the eyes (see Figure 20). The parietal lobes, on the upper rear of each hemisphere, handle judgments of weight, size, shape, and feel. The temporal lobes, near the temples, handle language and the perception of sound, among other things, which is not surprising given their proximity to the ears. And the frontal lobes, immediately behind the forehead, govern voluntary movements and some logical processes, which

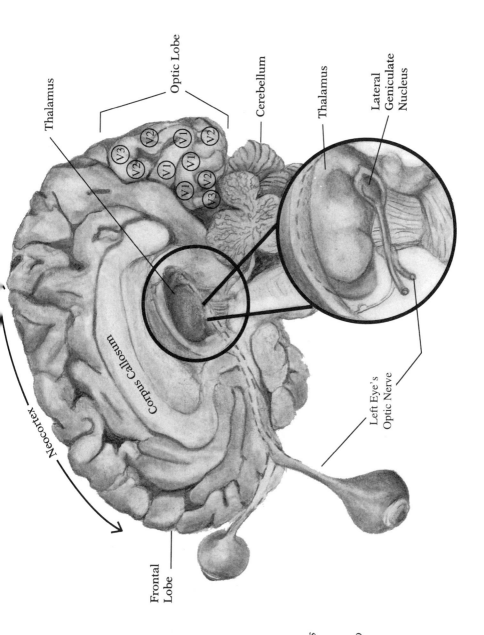

Figure 20. The right half of the human brain seen from the inside. The corpus callosum is a bundle of some 200 million nerve fibers that connects the two hemispheres. The lateral geniculate nuclei are way stations between the optic nerves and the optic lobes.

Thalamus

Optic Lobe

Cerebellum

Thalamus

Lateral Geniculate Nucleus

Left Eye's Optic Nerve

Neocortex

Corpus Callosum

Frontal Lobe

V3 V2 V1 V2 V2 V1 V1 V1 V2 V3

is why destruction to one from a lobotomy alters the personality. On such matters all neuroscientists agree.

But below this grand geographic representation of the brain's real estate, difficult arguments over jurisdiction begin. To understand why, let us consider what may appear to be a simple memory: that of your grandmother. Most of us are conscious of our grandmothers. But how? What series of neural links, of connections to connections, allow us to conjure up those dear old dames?

Presumably an image of grandma may come to mind, and that would appear to involve the optic lobes. But other parts of the memory relating to voice would have originated, presumably, in the temporal lobes. These recollections are connected to others related to things she said and did, the way her house smelled on Thanksgiving Day, the colors of her kitchen, and so on. Because the memory of your grandmother is no doubt imbued with emotional overtones, those cells, whose locations are not known, would also need to be activated. And finally there is the not inconsequential linguistic task of matching the word "grandmother" to one elderly or even long-deceased human female who happened to be the mother of one of your parents. It is difficult, if not impossible, to see how a single so-called "grandmother" cell would manage to bind all of these components of a complex memory together. This is known as the *binding problem*. Unlikely as it may seem, it is anything but simple, trivial, or mechanical to remember your grandmother.

Neuroscientists are hardly in agreement about precisely how such recollections arise or how, out of all the information in the brain, the phrase "your grandmother" engages so many different sections of the organ. And one of the major obstacles impeding the solution of the binding problem as applied to grandmothers is, fortunately, that no researcher anywhere in the world would perform experiments on equally beloved *grandchildren*. Ethical considerations preclude experiments that could, from a theoretical perspective, begin to resolve some of the important questions.

Beyond that, there are practical considerations. During behavioral experiments with the brains of freely moving rats, work conducted recently by Matthew Wilson and Bruce McNaughton at the University of Arizona, the maximum number of nerve cells that it has been possible to monitor at one time is approximately 150 (Wilson and McNaughton, 1993). This is, on its own terms, an impressive figure, but even in a *rat* brain we should not delude ourselves into thinking we have done anything more than keep tabs on a fraction of a fraction of the intellect of the creature. Its rodent memories of its own grandmother, for now, seem safe.

LAYERED STRUCTURES

Let's take a brief look at early brain theory. In their 1943 paper, McCulloch and Pitts attempted no less than to work out a neural basis for memory, and to this end they distinguished between theories of neural "nets with circles"—which have closed causal paths for logical feedback—and those without. Since "nets without circles" lack the capacity for reverberatory activity or sustained patterns of firing, they can be analyzed with much less difficulty, a distinction that continues to influence research on the dynamics of neural network models.

Because the theory of neural nets without closed causal loops is simpler to analyze, many neuroscientists begin with it, and some never get beyond it. In this theory, a nerve net is viewed as a *layered structure* in which the input layer (A) receives signals from a *retina* (R) that is connected to the outside world, perhaps optically. The input layer performs computations on the retinal data and sends the results of these computations to an intermediate layer (B). This intermediate layer performs further computations and sends the results to an output layer (C). Thus:

$$R \to A \to B \to C .$$

The key feature of this design is that the chain of logical consequences goes in only one direction: from the input layer toward the output layer. Never from the output layer to the input. Nowhere does the chain of consequences loop back onto itself, like the uroboros of antiquity, a mythical monster that eats its own tail: thus "nets without circles."

According to the theory, a particular neuron in the output layer attempts to decide whether the pattern on the retina is a house, a dog, the enemy, something to eat, your grandmother, or whatever. Such a neuron is called a "grandmother cell," and the general picture is in accord with the ideas of *stimulus-response* theory, which dominated much of psychology during the first half of the century.

THE PERCEPTRON

In 1965, McCulloch wrote that his famous paper with Pitts "might have remained unknown" but for the interest taken in it by physicist John von Neumann in connection with von Neumann's postwar lectures on the theory of computing machines (McCulloch, 1965). This display of modesty was both uncharacteristic of McCulloch and inappropriate, because "A logical calculus of the ideas immanent in nervous activity" provided the theoretical basis for neural network theory during the 1950s and beyond.

One focus of this early activity was the Perceptron research, which began at Cornell University as an attempt to construct computing machines that mimic brains (Block, 1962; Rosenblatt, 1958). Another, called ADALINE (an acronym for ADAptive LINear Element), was developed at Stanford University to improve the computational reliability of control systems in engineering applications (Widrow and Angell, 1962). Both of these systems are layered structures of McCulloch-Pitts model neurons, with one extra feature. The parameters of each model neuron (the input weightings and the threshold level) are adjusted according to some training rule that is based on past performance. A number of theoretical and experimental studies showed that it is possible to improve the system's ability to assign categories to patterns and obtain a rudimentary learning machine (Nilsson, 1965).

Such systems have an interesting design limitation: the number of distinct patterns that can be recognized is about equal to the number of neural elements. To an engineer, this suggests that one might as well categorize the patterns of interest by templates (one template shaped like the letter A to recognize that letter, for example) and devote one grandmother neuron to each pattern. This would eliminate a need for the training period and reduce the system to a coding device.

In 1969 computer scientists Marvin Minsky and Seymour Papert published a careful study of what Perceptrons *cannot* do. It is difficult, for example, for a Perceptron to recognize a pattern that is surrounded by a busy background, whereas human brains can do this rather well. This criticism stifled research in the area for more than a decade. A recent paper by John Hopfield (1982)—and the fact that Perceptrons can now be made with many thousands of elements—revived interest in this area of research among physicists at the beginning of the 1980s.

Layered learning machines that are based on the input-output principle of the Perceptron have many important applications—fingerprint detection, sorting and analysis of satellite data, searching for abnormal cells in tissue photographs, to name a few—but such activities are of marginal interest in understanding the reverberatory dynamics of the neocortex. Thus we shall concentrate our attention on "nets with circles."

THE CELL ASSEMBLY

Nets without circles are far removed from the "enchanted loom" that was described by Sherrington, as McCulloch and Pitts were well aware; thus they also gave careful consideration to nets with circles (McCulloch, 1947, 1965; Pitts and McCulloch, 1947). Progress was slow, however, for such nets are exceedingly difficult to analyze in detail. Why is this so? Because there are so many possible ways to arrange and interconnect the neurons of a brain.

The number of possible brains (each composed of ten billion McCulloch-Pitts neurons with inputs from ten thousand other neurons) is calculated in Appendix H as about

$$10^{10^{17}}.$$

This is a very large number. To get some idea how large it is, suppose it to be written out in the form:

$$1,000,000,\cdots,000,000.$$

Merely to write $10^{10^{17}}$ in this way would require some two hundred billion books the size of this one. Expressed in terms of the immense number $\Im (= 10^{110})$, which was defined in Chapter 2, the number of possible brains is greater than

$$\Im^{10^{16}}.$$

This is the immense number multiplied by itself ten thousand trillion times. Perhaps it should be called *hyperimmense*. How might such a vast system of neurons be organized?

An answer to this question was provided in Donald Hebb's book *The organization of behavior*. To appreciate the context in which this book appeared, note that midcentury North American psychology was divided among the claims of three different schools of thought. The first of these was Gestalt psychology, which regarded the neocortex of the brain as a homogeneous mass governed by a *field theory* (something like Maxwell's electromagnetism) that generated global patterns for motor control. The second school was stimulation-response connectionism, which viewed the brain as a Perceptron-like switchboard that makes connections between input and output nerve fibers. Finally came the behaviorists, who sought to make psychological theory more scientific by divorcing it from its physiological roots (Klein, 1980).

Perception was the purview of Gestalt theorists, while learning was dominated by connectionism and behaviorism, but none was particularly successful in accounting for all aspects of the real experimental data.

It was in this context that Hebb proposed a more interdisciplinary approach. His stated goal was "to bridge the gap between neurophysiology and psychology." He emphasized the importance of understanding the *neural* basis of thought, and to this end he introduced his central concept of the *cell assembly* as follows:

> Any frequently repeated, particular stimulation will lead to the slow development of a "cell-assembly," a diffuse structure comprising cells... capable of acting briefly as a closed system, delivering facilitation to other such systems and usually having a specific motor facilitation. A series of such events constitutes a "phase sequence"—the thought process. Each assembly may be aroused by a preceding assembly, by a sensory event, or—normally—by both. The central facilitation from one of these activities on the next is the prototype of "attention."

It is important to emphasize this statement because recent references to Hebb's work often overlook his seminal contributions to focus attention on his postulate that the strength of a synapse increases as it participates in firing a neuron, a property often called "Hebbian." Hebb was amused by this turn of events because the synaptic postulate was one of the few features of the theory that he did not consider to be original (Milner, 1993).

Hebb's first publication on the cell assembly was related to chimpanzees he had raised in a laboratory where, from birth, he could control their every stimulus. Such animals, he noted, exhibited spontaneous fear upon seeing a clay model of a chimpanzee's head (Hebb, 1946). The chimps, Hebb knew, had never seen a decapitation, yet some of them screamed, defecated, fled from their outer cages to the inner rooms where they were not within sight of the clay model; those that remained within sight stood at the back of the cage, their gaze fixed on the model held in my hand. (Hebb, 1980)

Such responses are clearly not reflexes; nor can they be explained as conditioned responses to stimuli, for there was no prior example in the animals' repertoire of responses. Similarly, they could have earned no behavioral rewards by acting in such a manner. But the actions of the chimps do make sense when described as disruptions of highly developed cell assemblies in which the chimps somehow understood the heads as decapitated representations of themselves.

Hebb's work (1949), accompanied by papers from McCulloch and Pitts (McCulloch, 1947; McCulloch and Pitts, 1943; Pitts and McCulloch, 1947), launched a generation of neuroscientists on a course of studying the dynamic properties of assemblies of interconnected neurons (Ashby et al., 1962; Caianiello, 1961; Rapoport, 1952; Rochester et al., 1956; Smith and Davidson, 1962; White, 1961). A salient fact that emerged from these studies was that a cell assembly will have a threshold level of activity for sustained firing. In other words, just as a single neuron requires some threshold number of input signals to its dendrites in order to fire through its axon, a cell assembly of neurons is activated by some typical level of excitation.

To see this, consider the following simplified model of a cell assembly. It is composed of McCulloch-Pitts neurons, each having excitatory inputs coming from random connections to the other neurons; for analytical convenience, time is assumed to be divided into discrete intervals of length τ. (This simplifying assumption is sometimes called *time quantization*, a term that has nothing to do with Schrödinger's quantum theory.) With reasonable parameters for the system, one finds that the probability of a neuron firing in the next value of time $(t + \tau)$ is a *sigmoid* (or ess-shaped) function (S) of the fraction (F) of neurons firing at time t. Such a function is sketched in Figure 21, where it can be seen that there are three values for the firing rate $(0, F_0,$ and 1) at which the probability of firing is just equal to the firing rate. These are equilibrium states of the system, but only two of them $(F = 0$ and $F = 1)$ are stable.

If the initial value of F is slightly less than F_0, then the firing rate goes to zero along the downward staircase in Figure 21. If the initial value of F is slightly greater than F_0, on the other hand, $F(t)$ goes to one along the upward staircase. Thus $F = F_0$ is an unstable firing rate for the system. In other words, if input signals cause a firing level that is less than F_0, the activity of the assembly will relax back to zero, but if the firing rate is brought above F_0, the activity will ignite and rise to its maximum level.

Just as a bonfire has a thermal threshold, so does the lighting of each single twig or branch. In the context of neurodynamics, this observation can be restated as follows:

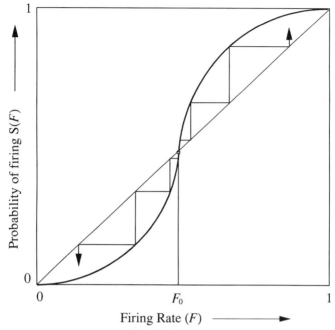

Figure 21. *A graph of the probability of firing at time t + τ for an assembly of randomly interconnected neurons as a function of the firing rate at time t. This figure shows the threshold behavior of an assembly of interconnected neurons.*

An assembly of randomly interconnected neurons has the same "all-or-nothing" and threshold behaviors as a single neuron.

To better understand what Hebb and others now mean by a cell assembly, it can be portrayed by way of another analogy: society. We can compare the brain to a city and the individual neurons to its citizens. A particular citizen might be a member of several social assemblies, such as a political association, a church, a bowling league, a parent-teacher group, a hiking club, the Junior League, and so on. The members of each social assembly are interconnected because, inevitably, each has a few overlapping members. What's more, the lists of addresses and telephone numbers allow one organization to activate its own members—or even the members of a like-minded assembly—whenever an appropriate occasion arises.

Members of the hiking club, for instance, could encourage the 4-H Club to resist the development of a theme park near their farms. Or the Junior League might enlist the support of teachers to close a pornographic bookstore. Just as an individual could be a member of both the hiking club and the League, a single nerve cell would—in Hebb's theory—participate in many different assemblies of the brain.

Here, for the first time, there is an inkling of a neural theory that could solve the problem of how we are aware of our grandmother: an assembly of cells, in different parts of the brain, could have coalesced over many years in response to a particular elderly matron. Cells in the optic lobes would respond to her image and tend to fire those cells in the temporal lobes that had previously been activated in response to her voice. Given sufficient time with one's grandmother, one's memory might be photographic, just as church leaders and the homeless might get to know each other well over long winter months, or they might never really be acquainted at all.

THE HIERARCHY OF CELL ASSEMBLIES

At this point, it is important to recognize that the concept of the cell assembly is essentially hierarchical in nature. Most of us can agree that protons and electrons are the stuff of atoms; that atoms are the stuff of molecules; that biological molecules, as we saw in Chapter 3, exist on a level of complexity all their own; and so too with nerve impulses and solitary neurons. Now, by the same token, thanks to Hebb and those who have followed in his footsteps, assemblies of neurons also seem to operate in ways that do not apply at any *other* levels of the hierarchy.

To understand the peculiarities of cell assemblies, consider how an infant might learn to perceive the triangle T shown in Figure 22a. The constituent sensations of the vertices are first supposed to be centered on the retina by eye movement and mapped onto the primary visual area (area V1) of the optical lobes, shown in Figure 20. Corresponding cell assemblies E, F, and G then develop in the secondary visual area (V2) through nontopological connections with area V1. The process of examining the triangle involves elementary phase sequences in which E, F, and G are sequentially ignited. Gradually these subassemblies will fuse into a common assembly for perception of the triangle T.

With further development of the assembly T—which reduces its threshold (F_0 in Figure 21) through the development and strengthening of the internal connections among E, F, and G—a glance at one corner, with a few peripheral clues, can quickly ignite the entire assembly representing T. At this point in the learning process, T is established as a higher-order cell assembly for perception of a triangle, and E, F, and G are its constituent subassemblies.

Since assemblies of neurons have the threshold behavior shown in Figure 21, assemblies of assemblies (and assemblies of assemblies of assemblies) are expected to share the same property. Each has its own minimal level of stimulation that is necessary for it to fire. Once that level is reached, however, the whole assembly ignites. This hierarchical ordering is in accord with what we know from the experience of learning to read. First one must learn to distinguish letters from lines and curves and angles, next words—which are assemblies of letters—must be learned, then phrases, then sentences, then paragraphs, and so on. Whereas the threshold for a first-order assembly

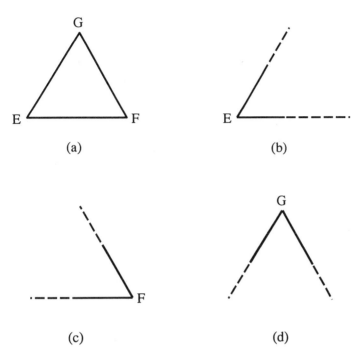

Figure 22. *Diagrams related to Hebb's discussion of learning to perceive a triangle T.*

is a certain number of active neurons, the threshold for a second-order assembly is a certain number of first-order assemblies, and the threshold for a third-order assembly is a certain number of second-order assemblies. Understanding the word "grandmother" thus involves a massive number of subassemblies: just try measuring all the angles in the letters, let alone every cell that single word engages. But such assemblies do concatenate rapidly within the brain, rippling from one to the next based upon all of their constituent subassemblies, and so we can quickly recall a particular face, a certain voice, a wonderful holiday fragrance.

Hebb first entertained the cell assembly theory in the mid-1940s after rereading Marius von Senden's *Space and sight*, which was originally published in Leipzig, Germany, in 1932. In this work von Senden assembled existing records on 65 patients from the eleventh century up to the year 1912 who had been born blind due to cataracts. At ages varying from 3 to 46, the cataracts were removed, and a variety of reporters observed them as they went about handling the sudden and often maddeningly novel influx of light.

One of the few uniformities in these cases, von Senden noted, was that the process of learning to see "is an enterprise fraught with innumerable difficulties, and that the common idea that the patient must necessarily be delighted with the gifts of light and colour bequeathed to him by the operation is wholly remote from the facts." Not every patient rejoiced upon being forced to make sense of light that was all but incomprehensible, and many found the effort of learning to see so difficult that they simply gave up. The doctor of a twenty-one-year-old Viennese woman reported:

> Even before the operation she was dull and quite literally spirit-less, as I have never yet seen such a patient before. Whenever I visited her afterwards, the bandages having been long since discarded, I found her with her eyes closed; though not forced to do so by any aversion to the light; and laborious persuasion was needed before she would even come to look at the things immediately about her, and at last to become acquainted with them. Some years ago, indeed, her unfortunate father, who had hoped for so much from this operation, wrote that his daughter carefully shuts her eyes whenever she wishes to go about the house, especially when she comes to a staircase, and that she is never happier or more at ease than when, by closing her eyelids, she relapses into her former state of total blindness.

If seeing were a simple reflex, as the stimulation-response theorists held, the restoration of vision should have completed a circuit, allowing a patient to view the world as lucidly as if he or she had done so all along. Instead von Senden's accounts suggest that seeing was not at all automatic. The sweet-heart of one fifteen-year-old patient from southern France, in an attempt to encourage him to learn to see, tried the following experiment:

> One day, during the vine-harvest, she picked a bunch of grapes and showed it to her lover from a distance.
>
> "What is that?" she asked him.
>
> "It is dark and shiny," he replied.
>
> "Anything else?"
>
> "It isn't smooth, it has bumps and hollows."
>
> "Can one eat it?" she asked.
>
> "I don't know."
>
> "Then take it and try."
>
> As soon as he touched the bunch, he cried: "But they're grapes!

The reports indicated that patients who had been completely dependent on tactile impressions before the operation had an awareness of space (if it can be so called) that was totally different from a normal visual awareness. At first, von Senden discovered, "the patient feels visual impressions to be something alien, intruding on his mind without action on his own part."

Later, he reported, "the stimuli impinging on the visual organ from an objective shape merely occasion the act of perception as such, but do not determine its outcome... The final development up to the fully formed idea of shape involves a series of transformational forms as intermediate stages that develop one from another and are liable to vary between individuals, since it is the individual who creates them." A nine-year-old boy from Leeds, for example, spent days trying to learn how to tell a sphere from a cube. From his record:

> He gradually became more correct in his perception, but it was only after several days that he could tell by his eyes alone, which was the sphere and which the cube. When asked, he always, before answering, wished to take both into his hands. Even when this was allowed, when immediately afterwards the objects were placed before the eyes, he was not certain of the figure.

That such observations are not artifacts of the surgery or uniquely human was established through fortuitous observations on a pair of young chimpanzees that had been reared in the dark by Austin Riesen, a colleague of Hebb (Riesen, 1947). After being brought out into the light, these animals showed no emotional reactions to their new experiences. They seemed unaware of the stimulation of light and did not try to explore visual objects by touch. Hebb concluded that the chimps showed no visual response because they had not formed the assemblies that are necessary for visual perception.

If one accepts the concept of cell assemblies (Braitenberg, 1978), it is interesting to ask how many might be able to form (Scott, 1977; Palm, 1981, 1993). This is a difficult question, but Charles Legéndy (1967, 1975) has obtained some results from a simple model. He assumes that the brain is already organized into subassemblies and discusses their organization into larger assemblies. The assembly and one of its subassemblies variously represent "a setting and a person who is part of it, a word and one of its letters, an object and one of its details."

In Legéndy's model, interconnections are assumed to be evenly distributed over the neocortex to avoid the complications of spatial organization. His subassemblies are formed through *weak* contacts and assemblies emerge from subassemblies through the development of *latent* into *strong* contacts between neurons. From statistical arguments, Legéndy estimates the maximum number of assemblies in a brain to be about $(N/ny)^2$, where N is the number of neurons in the brain, n is the number of neurons in a subassembly, and y is the number of subassemblies in an assembly. Taking $N = 10^{10}$, $n = 10^4$, and $y = 30$, he finds

$$\sim 10^9 \quad \text{assemblies}$$

as a conservative estimate for "the number of elementary things the brain can know."

Hopfield has studied the emergent properties of a network of N Mcculloch-Pitts neurons and shown that memories can be stored as stable entities and recalled from excitation of sufficiently large subparts, which is just the threshold effect that is illustrated in Figure 21. Interestingly, he finds that the number of stable patterns that can be stored before recall error rises is about $.15N$. For $N = 10^{10}$, this is

$$\sim 1.5 \times 10^9 \quad \text{patterns},$$

a value that is in remarkable agreement with Legéndy's estimate.

The number of neurons in the neocortex may be an order of magnitude greater than 10^{10}, and the multiplex neuron of Figure 19 is much more complex than the simple nerve models assumed by Legéndy and Hopfield, but—as Legéndy points out—10^9 is approximately the number of seconds in 30 years. Thus a simple model of the brain—based on the McCulloch-Pitts neuron—provides sufficient storage for the complex memories of a normal lifetime.

It is important to notice that these conservative estimates for the number of complex assemblies in a human brain are much larger than the result one gets by dividing the number of neurons in the brain by the number of neurons in an assembly. As was suggested above in the context of the social assembly, each neuron can participate in many cell assemblies, thus the difference.

Since these estimates for the maximum number of assemblies in the brain are about equal to the number of neurons in the brain, one might wonder what is the advantage of a hierarchical organization of cell assemblies. Why does not the brain simply assign one "grandmother cell" in the neocortex to each concept that is to be perceived? The answers to this question are twofold. First, the organization of memory into a hierarchy of cell assemblies leads to *robust* storage of information. In other words, our species evolved with a certain neurological plasticity: injuries to cell assemblies to a temporal lobe, say, mean that one might tend to forget the *sound* of an advancing lion but that other cell assemblies would become ignited and trigger responses based on visual, olfactory, or other cues from the environment.

Such matters were not lost on Hebb. "How," he asked, "can it be possible for a man to have an IQ of 160 or higher, after a prefrontal lobe has been removed?" If a particular memory is stored in a certain neuron, its continued existence depends on the life of the cell. Cell death, in this model, means *total* loss of the memory. Assembly storage, on the other hand, is robust because physical damage—leading to the death of a few cells—will not necessarily destroy specific assemblies but instead degrade many by roughly equal degrees. This loss could then be restored through recruitment of additional cells into the assembly.

To return to the analogy with society, suppose that half of the members of the bowling league were killed in a bus accident. This wouldn't be the end of the organization; the surviving members could recruit additional bowlers until the original membership level was re-established. A damaged cell assembly would do the same, as we have all experienced with reinforcement of an almost forgotten memory, or seen in connection with recovery from a stroke. The brain can rebuild itself.

What about the binding problem? The remembering of one's grandmother brings together many disparate images: memories of her face that must reside in the occipital (visual) lobes and those in the temporal (auditory) lobes that recall her voice. These images are connected to others related to things she said and did, the way her house smelled on a holiday, the colors of her kitchen, and so on. It is difficult, if not impossible, to see how a single grandmother cell would manage to bind all of these components of a complex memory together. This is called the binding problem, and Hebb's cell assembly theory suggests a solution. As is shown in Figure 23, assembly A_1 might represent a visual image of grandmother's face and A_2 the layout of her kitchen. Assemblies B_1 and B_2, on the other hand, might recall the sound of her voice and something she once said. Both of these pairs of memories might be linked to assembly C, representing the smell of her creamed onions. Just as with the learning to see a triangle in Figure 21, the sharing and reinforcement of connections to C could lead to the formation of a complex assembly that includes all the features—A_1, A_2, B_1, B_2, and C—in one complex perception of Thanksgiving Day.

Although this appears to be a reasonable solution to the binding problem, the problem has continued to vex both psychologists and neurobiologists over the past four decades. A perceptive review of these discussions has

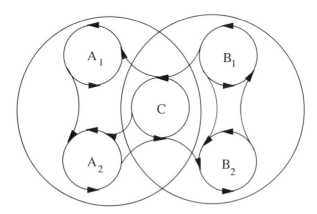

Figure 23. Hebb's solution for the binding problem. *A subassembly, C, may act as a link between two assemblies or conceptual complexes. One concept is represented by* A_1, A_2, *and C; the second by* B_1, B_2, *and C. The common subassembly, C, provides a basis for binding of the two assemblies. (Redrawn from Hebb, 1949.)*

recently been published by Valerie Hardcastle (1994), who notes that phase-locked oscillations may play a role in coupling the activities of individual neurons. But this is not enough. In her words:

> Since what we see on the larger scale reflects the statistical aggregation of underlying stochastic microprocesses, charting individual cellular behaviour might be analogous to following the activity of a single particle in a Bénard cell [a hydrodynamic experiment] in order to understand convection.

In contrast to traditional computer metaphors, which act on a single level of analysis, she proposes that bound perceptions should be tied to "particular activated *assemblies* of neurons," each with a multilevel system of organization. If we bear in mind the fact that two assemblies that are similar on the average can represent quite different codes, this conclusion is close to Hebb's fundamental description of the brain's activity.

Although the total number of neurons in a brain is certainly an important measure of its ultimate learning ability, Hebb emphasized the importance of the ratio of the total association area of the neocortex to the total sensory area (the A/S ratio) in determining the nature of learning. For lower creatures, such as worms and insects (and, of course, currently available computers), the A/S ratio is essentially zero, but with higher species (rat, dog or cat, monkey, chimpanzee, human) the ratio grows. A larger A/S ratio means that fewer cortical neurons are under sensory control at the start of the learning experience; thus a longer learning period is needed for the uncontrolled neurons to become organized. Since our vast repertoire ($\sim 10^9$ or more) of high-order assemblies must be established one-by-one over the formative years, initial learning by humans is relatively slow. In an adult, however, human learning is rapid because *within the context of the cell assembly repertoire that has developed*, a new situation can be interpreted and additional interactions develop quickly between established assemblies. The context of the established repertoire includes that of culture, which will be considered in Chapter 8.

THE CELL ASSEMBLY REVISITED

Hebb's classic work was a conservative account. In it he assumed nothing about the nature of neural systems that had not been confirmed by the electrophysiologists or neuroanatomists of his day (Hebb, 1949). In 1980, thirty-one years after his major book appeared, Hebb published a retrospective review of how the cell assembly concept arose and how the theory had fared. Although it was von Senden's (1932) observations of adult humans restored to vision, and their fortuitous confirmation in chimpanzees (Riesen, 1947), that converted Hebb to the cell assembly theory, this evidence has turned out to be less convincing than it had seemed at the time. In 1957,

Riesen and his colleagues found that raising chimpanzees in total darkness caused some of their retinal cells to alter their protein content and die (Chow et al., 1957), although no such effect was observed in the rats studied earlier by Hebb. This effect might have contributed to the difficulties in learning to see that the chimps experienced. In 1963, it was reported that kittens deprived of patterned stimulation lost connections in the lateral geniculate bodies (see Figure 20) (Wiesel and Hubel, 1963). Since either or both of these effects may have occurred in the patients observed by von Senden, Hebb said that he "might well have dropped the whole thing" if he had known of them in the 1940s. But at the time he and his colleagues were encouraged to begin a program of research to test the cell assembly hypothesis.

According to the theory, adult thought processes involve the activity of cell assemblies, which in turn are organized by sensory stimulation during the learning period of a young animal. Confirmation for the theory was provided by experiments showing that rats reared in a rich perceptual environment (a "Coney Island for rats") were much more intelligent as adults than those raised in a restricted environment (Rabinovitch and Rosvold, 1951). This effect occurred only during youthful development, not during adulthood. Similar experiments with Scottish terriers showed even more striking results, as is expected from the cell assembly theory because the A/S ratio is larger for a dog than for a rat (Thompson and Heron, 1954). Terriers reared in single cages where they could not see or touch other dogs had abnormal personalities and could not be trained or bred. Other studies showed that dogs reared in restricted environments did not respond to pain, as if they were lobotomized (Melzak and Scott, 1957).

It should be noted that these experiments on learning in laboratory animals are of considerably more than theoretical interest to professional psychologists. In 1940, Beth Wellman and her colleagues were being ridiculed for suggesting that richer living environments could increase the intelligence levels of deprived children (Wellman, 1940). The animal experiments conducted by Hebb and others give strong support to those who emphasize the importance of a variety of experiences during the formative years.

Monotony also causes difficulties in the short term. In his original formulation of the cell assembly theory, Hebb (1949) expected that perceptual isolation would create problems because the phase sequence needs meaningful sensory stimulation to remain organized. To test this aspect of the theory, experiments on perceptual isolation were performed by Woodburn Heron and his colleagues in the mid-1950s (Heron, 1957). In these experiments the subjects were college students who were paid to do nothing. Each subject lay quietly on a comfortable bed wearing cardboard arm cuffs and translucent goggles and heard only a constant buzzing sound for several days. During breaks for meals and the toilet, the subjects continued to wear their goggles, so they averaged 22 hours a day in total restriction.

Many subjects took part in the experiment intending to plan future work or prepare for examinations. According to Hebb (1980), the main results were that the subjects' ability to solve problems in their heads declined rapidly after the first day as it became increasingly difficult to maintain coherent thought, and for some it became hard or impossible to daydream. On the third day, roughly, hallucinations became increasingly complex. One student said that his mind seemed to be hovering over his body in the cubicle, like a ball of cotton wool. Another reported that he had two bodies but didn't know which was really his. Such observations support a wide variety of anecdotal reports from truck drivers, shipwreck survivors, solitary sailors, long-distance drivers, and the like that extended periods of monotony breed hallucinations. (During his famous flight across the Atlantic Ocean, for example, Charles Lindbergh [1953] reported "vapor-like shapes crowding the fuselage, speaking with human voices, giving me advice and important messages.")

After these perceptual isolation experiments were concluded, the subjects reported difficulties with visual perception that lasted for several hours and were found to have a 10% slowing of their electroencephalograms or brain waves. They were also more vulnerable to propaganda.

Although the specific results of such experiments were not predicted by the cell assembly theory, the disorganizing effect on thought had been anticipated. In general, isolation experiments with rats, dogs, and humans—all with increasing A/S ratios—tend to support the cell assembly theory, but they also indicate that there is much to learn about how the assemblies interact and function.

In Hebb's view, some of the strongest evidence in support of the cell assembly theory was obtained from *stabilized-image* experiments, which were arranged as indicated in Figure 24 (Hebb, 1980; Milner, 1993; Pritchard et al., 1960; Pritchard, 1961). A triangle or a square was projected as a fixed image onto the retina, and the subjects were asked to report what they saw. Since this is an introspective experiment, typical results are displayed in a thought balloon. Many subjects, including Hebb himself, reported that the component lines (subassemblies) of a triangle and a square would jump in and out of perception all at once, exactly as is expected from the original formulation of the theory and the learning sequence indicated in Figure 22. Others perceived partial losses of the elements of the figures, showing that an assembly representing a simple geometrical figure can be composed of a rather large number of constituent subassemblies.

Hebb also notes that support for the cell assembly theory is provided by the physiological observations of David Hubel and Torsten Wiesel, who showed in 1962 that the cat's visual cortex contains a hierarchy of neurons. *Simple cells* are first excited by specific lines in the visual field. Then *complex cells* are excited by the simple cells. Finally *hypercomplex cells* are excited by complex cells. This is a hierarchical structure of the form

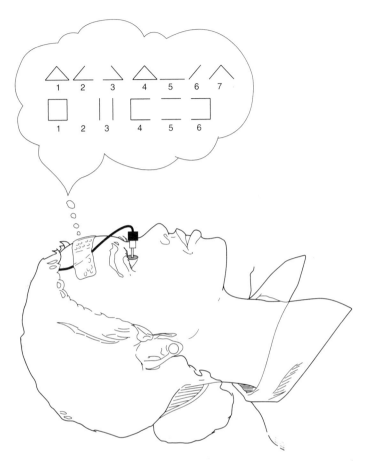

Figure 24. *Sketch of contact lens and optical apparatus mounted on the eyeball of a reclining observer. The wire is connected to a small lamp that illuminates the target. The thought balloon shows sample sequences of patterns perceived by the subject with images that are stabilized on the retina by the apparatus. In the upper (lower) row a triangle (square) is the target. (After a photograph in Pritchard, 1961.)*

Hypercomplex cells
↑ ↓
Complex cells
↑ ↓
Simple cells

which is qualitatively similar to the hyperstructure of biochemical activity that was proposed by Eigen and Schuster (1979) (and discussed in Chapter 3) as the basis for life. Hebb found the parallel of simple, complex, and hypercomplex cells with first-, second-, and third-order cell assemblies to be obvious, making the existence of higher-order cell assemblies a reasonable speculation.

Donald Hebb was a careful scientist. In his original formulation of the cell assembly theory, he assumed only excitatory interactions between neurons, in accord with the prevailing opinion in electrophysiology at the time. In 1952, however, this restriction was relaxed by observations of inhibitory interactions reported by John Eccles and his colleagues (Brock, et al., 1952), which led Peter Milner (1957) to propose refinements of the theory. In the context of Figure 21, inhibitory cortical neurons allow several intermediate stable states of activity lying between no neurons firing and all neurons firing. (The single sigmoid curve is replaced by a wiggly curve that crosses over the $S = F$ line in several places.) From a global perspective, this enhances the opportunities for one assembly to be ignited without interference from others.

Finally we turn to a question that puzzled Hebb for many years: What is the relationship between brain size and intelligence? Whales and elephants have brains that are three to five times larger than those of humans, yet we can outwit them. Although there is some reason to suppose that the ratio of brain weight to body weight might be the important factor, there are many exceptions. The average human brain weighs about 1300 grams, but eminent persons have had brains of 1100 grams and dullards often have much larger ones (Hobb, 1980). Above a certain minimum size, what other feature is important? Why is the substance of the human brain so superior to those of our near relatives?

During a discussion of the role of cortical inhibition in shutting down assemblies after they fire, Milner called Hebb's attention to measurements of the relative fractions of inhibitory neurons in various mammals (Rakic, 1975). These marked differences are displayed in Table 1. Hebb suggested that the inhibitory interneurons may have the function of "streamlining the thought process" by reducing the time required for an assembly to relax back into its quiescent (latent) state after having been ignited. The shorter this time becomes, the faster the shuttle can move about the enchanted loom.

Species	Inhibitory neurons
rabbit	31%
cat	35%
monkey	45%
human	75%

Table 1. The fraction of inhibitory neurons in various mammals.

THE NATURE OF THOUGHT

Hebb's approach to the nature of brain dynamics, and also that of Mc-Culloch and Pitts, is *bottom up*. They attempt to account for the facts of psychology using the facts of neurology, but the track through the neural jungle is soon lost. To appreciate what the neuroscientist is up against, imagine trying to determine the details of cell dynamics from the laws of chemistry without being able to make experimental observations of proteins, DNA, biochemical reactions, mitochondrial structure, and so on. Although it seems clear that the brain is organized in a hierarchical structure with several layers of dynamic activity, the details of this organization are exceedingly difficult to map out.

Thus another approach is *top down*: to understand the dynamic nature of thought through introspection. Since these two approaches are studying the brain from opposite sides of the forest, so to say, both can be valuable.

As an example of a well-controlled introspective experiment, consider the jumping back and forth between two perceptions of an ambiguous figure. The first record of such an observation is in an 1832 paper by Louis Necker, a Swiss geologist. Upon viewing the representation of the cube shown in Figure 25a, Necker noted "a sudden and involuntary change" in its orientation; first one face seems to be in front and the other behind, then the impression is reversed. (Try it yourself. How rapidly can your brain switch back and forth between the two perceptions?) Another early example is the Schröder stairs, shown in Figure 25b, which appeared in the chemistry literature in 1858 (Schröder, 1858).

Such effects have long fascinated the Gestalt psychologists, who see the perceptions as global states of a brain field, but they are also readily understood in the context of Hebb's theory. In this theory there is an assembly for each of the two perceptions, but attentiveness to one cannot be sustained—because of habituation or tiring of the active neurons—so ignition jumps from one to the other every second or two. In the jargon of computer engineering, the two assemblies or perceptions act as a *flip-flop* circuit.

Many examples of ambiguous perception appear in the world of visual art. Well known are *Slave market with apparition of the invisible bust of Voltaire* by Salvador Dali (Atwater, 1971) and Pablo Picasso's somewhat more subtle *Marie Thérèse Walter*, who appears either in front view or profile depending upon the mental state of the observer (Scott, 1977). The Dutch artist Maurits C. Escher was particularly fond of such effects. Roger Penrose, an English mathematical physicist, and his colleagues have recently compiled a fascinating collection of studies that discuss the relationships between Escher's work and the concepts of visual perception, mathematics and physics (Coexter et al., 1986).

Literature, too, exhibits many instances of ambiguous perception. The Japanese *haiku* requires of its seventeen syllables—among references to the seasons and to nature—an unexpected shift of mood. In English literature, especially poetry and drama, William Empson (1947) has described seven different sorts of ambiguity that range from extra meanings of a phrase to

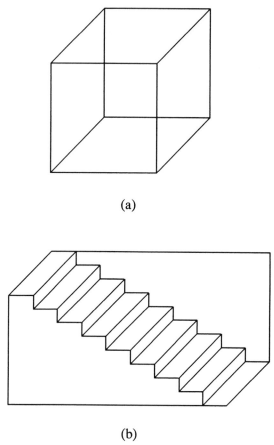

(a)

(b)

Figure 25. (a) The Necker cube. (b) The Schröder stairs.

evidence of contradiction in the mind of the author. Not surprisingly, the works of William Shakespeare are well represented, and his tragic drama *Hamlet* especially so.

As but a single example from Empson's book, consider poor Ophelia. Her father has been killed by Hamlet and his father—the rightful King—has been killed by a false brother in league with Hamlet's mother. When Ophelia enters—bereft, deserted, and deranged—and asks:

Where is the beauteous Majesty of Denmark?

who can know what she means? We can be sure that Shakespeare was aware of the many possibilities and that he encouraged them. We can almost hear the shuttling of *his* enchanted loom.

Finally, Hebb's concept of the phase sequence, in which one thought leads to another and then another under the guidance of external stimulation, is closely related to the stream of consciousness so effectively rendered by James Joyce (1934):

> He crossed to the bright side, avoiding the loose cellarflap of number seventyfive. The sun was nearing the steeple of George's church. Be a warm day I fancy. Specially in these black clothes feel it more. Black conducts, reflects (refracts is it?), the heat. But I couldn't go in that light suit. Make a picnic of it. His eyelids sank quietly often as he walked in happy warmth. Boland's breadvan delivering with trays our daily but she prefers yesterday's loaves turnovers crisp crowns hot. Makes you feel young. Somewhere in the east: early morning: set off at dawn, travel round in front of the sun, steal a day's march on him. Keep it up for ever never grow a day older technically. Walk along a strand, strange land, come to a city gate, sentry there, old ranker too, old Tweedy's big moustaches leaning on a long kind of a spear. Wander through awned streets. Turbaned faces going by. Dark caves of carpet shops, big man, Turko the terrible, seated crosslegged smoking a coiled pipe. Cries of sellers in the streets. Drink water scented with fennel, sherbet. Wander along all day. Might meet a robber or two. Well, meet him.

CORTICAL STIMULATION

Those who find such artistic and literary observations too subjective to be scientifically meaningful should consider the hallucinations induced by the Canadian neurosurgeon Wilder Penfield and his colleagues through electrical stimulation of the neocortex (Penfield and Perot, 1963). The purpose of such stimulation was to locate the origin of epileptic activity in order to remove the offending portion of the cortex. In this study, the records of 1288 brain operations for focal epilepsy were examined. Gentle electrical stimulation was applied to the temporal lobes of 520 patients, of whom 40 reported experiential responses. Stimulating currents between 50 and 500 microamperes were used in pulses of 2 to 5 milliseconds at frequencies of 40 to 100 hertz.

M.M. was a typical case. A woman of 26, she had her first epileptic seizure at the age of five. When she was in college, the pattern included hallucinations, both visual and auditory, coming in "flashes" that she felt she had experienced before. One of these

> had to do with her cousin's house or the trip there—a trip she had not made for ten to fifteen years but used to make often as a child. She is in a motor car which had stopped before a railway crossing. The details are vivid. She can see the swinging light at the crossing. The train is going by—it is pulled by a locomotive passing from the left to right and she sees coal smoke coming out

of the engine and flowing back over the train. On her right there is a big chemical plant and she remembers smelling the odor of the chemical plant.

During the operation, her skull was opened and her right temporal region explored to locate the epileptic region. Figure 26 shows the exposed temporal lobe with numbered tickets that mark the sites at which positive responses were evoked. Penfield termed the responses *experiential* because the patient felt that she was reliving the experience—not merely remembering it—even as she remained aware that she was lying in the operating room and talking to the doctor.

The following experiential responses were recorded upon electrical stimulation at the numbered locations:

11. She said, "I heard something familiar, I do not know what it was."

11. Repeated without warning. "Yes, sir, I think I heard a mother calling her little boy somewhere. It seemed to be something that happened years ago." When asked if she knew who it was she said, "Somebody in the neighbourhood where I live." When asked she said it seemed as though she was somewhere close enough to hear.

11. Repeated 18 minutes later. "Yes, I heard the same familiar sounds, it seems to be a woman calling. The same lady. That was not in the neighbourhood. It seemed to be at the lumber yard."

13. "Yes, I heard voices down along the river somewhere—a man's voice and a woman's voice calling." When asked how she could tell it was down along the river, she said, "I think I saw the river." When asked what river, she said, "I do not know, It seems to be one I was visiting when I was a child."

13. Repeated without warning. "Yes, I hear voices, it is late at night, around the carnival somewhere—some sort of a travelling circus. When asked what she saw, she said, "I just saw lots of big wagons that they use to haul animals in."

12. Stimulation without warning. She said, "I seemed to hear little voices then. The voices of people calling from building to building somewhere. I do not know where it is but it seems very familiar to me. I cannot see the buildings now, but they seemed to be run-down buildings."

14. "I heard voices. My whole body seemed to be moving back and forth, particularly my head."

14. Repeated. "I heard voices."

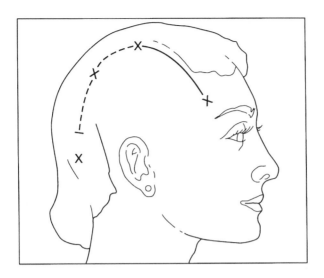

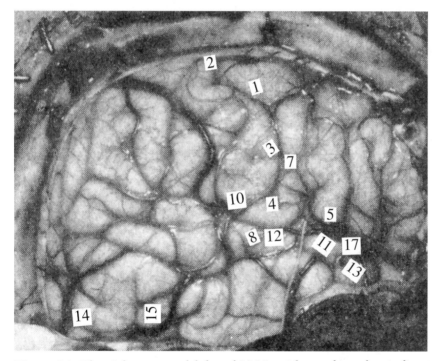

Figure 26. *The right temporal lobe of M.M. with numbers that indicate points of positive stimulation. (Redrawn from Penfield and Perot, 1963.)*

The forty cases surveyed by Penfield and Perot show that this particular example of experiential response is not at all unique. They conclude that, in the human brain, there is a remarkable record of the stream of each individual's awareness or consciousness and that stimulation of certain areas of the neocortex—lying on the temporal lobe between the auditory and visual sensory areas—causes previous experience to return to the mind of a conscious person.

A DEEPER QUESTION

To recall our flickering candle of Chapter 4, such gentle electrical current seems to be a match that lights bonfires in the mind. But these are magic bonfires. A billion or more of them rest in a normal head, each intertwined with all the others. When the match is applied, one lights, then another, then another. They do not all burn at once. From a few sparks, it seems, the brain can effortlessly elicit a record of the stream of awareness. In our electronic age, it is common to immediately ask whether the interactions between these magic bonfires can be modeled on a digital computer.

Before attempting an answer, one should consider the following points. First, the caveat "in principle" is necessary, because there is no hope of making a model of all neural interconnections with currently available computers. And as was noted at the end of the previous chapter, we are, on top of everything, uncertain about the dynamic nature of the individual neurons.

Finally, even assuming the McCulloch-Pitts representation of a neuron is roughly correct (which it is *not*), one is beset with ignorance about the cell-assembly dynamics. Some progress has been made in tracing the interconnections between individual neurons in the visual cortex (Eccles, 1973; Hubel and Wiesel, 1962), but—from the perspective of electrophysiology—almost nothing is known about the ways that complex assemblies might interact. Word association tests, introspection, dream analysis, and psychotherapy may provide some additional information, but since each subject is a member of a hyperimmense set—the set of all possible people—it is difficult to make useful generalizations.

Beyond the practical problem of whether we can model the enchanted loom and how we might do so, there is another question. A deeper question. How can science describe the experience of seeing? In his Gifford Lectures Charles Sherrington (1951) put it thus:

> A star we perceive. The energy scheme deals with it, describes the passing of radiation thence into the eye, the little light-image of it formed at the bottom of the eye, the ensuing photo-chemical action of the retina, the trains of action potentials travelling along the nerve to the brain, the further electrical disturbance in the brain, the action-potentials streaming thence to the muscles of eye-balls and of the pupil, the contraction of them sharpening under the light-image and placing the seeing part of the retina

under it. The "seeing"? That is where the energy-scheme forsakes us. It tells us nothing of any "seeing". Much, but not that.

Throughout the past century, this chasm between the details of a mechanistic explanation of the brain and the ever-present reality of conscious awareness, has continued to yawn. Whatever mechanistic explanation one might construct to explain the nature of the mind, one can as well imagine that same mechanism working *without* the feeling. Reductive materialism fails to bridge the gap.

References

W R Ashby, H von Foerster, and C C Walker. Instability of pulse activity in a net with threshold. *Nature*, 196:561–562, 1962.

F Atwater. Multistability in perception. *Sci. Am.*, 225:62–71, December 1971.

V Braitenberg. Cell assemblies in the cerebral cortex. In *Theoretical approaches to complex systems*, Springer-Verlag, Berlin, 1978.

H D Block. A model for brain functioning. *Rev. Mod. Phys.*, 34:123–135, 1962.

L G Brock, J S Coombs, J C Eccles, R Heim, and G Palm, editors. The recording of potentials from motoneurons with an intracellular electrode. *J. Neurophysiol.*, 117:431–460, 1952.

E R Caianiello. Outline of a theory of thought-processes and thinking machines. *J. Theoret. Biol.*, 1:204–235, 1961.

H S M Coexter, M Emmer, R Penrose, and M L Teuber. *M.C. Escher, art and science*. North-Holland, New York, 1986.

K L Chow, A H Riesen, and F W Newell. Degeneration of retinal ganglion cells in infant chimpanzees reared in darkness. *J. Comp. Neur.*, 107:27–42, 1957.

J C Eccles. *The understanding of the brain*. McGraw-Hill, New York, 1973.

M Eigen and P Schuster. *The hypercycle: A principle of natural self-organization*. Springer-Verlag, Berlin, 1979.

W Empson. *Seven types of ambiguity*. New Directions, New York, 1947.

V G Hardcastle. Psychology's binding problem and possible neurobiological solutions. *J. Consciousness Stud.*, 1:66–90, 1994.

D O Hebb. On the nature of fear. *Physiol. Rev.*, 53:259–276, 1946.

D O Hebb. *The organization of behavior*. Wiley, New York, 1949.

D O Hebb. *Essay on mind*. Lawrence Erlbaum Associates, Hillsdale, New Jersey, 1980.

D O Hebb. The structure of thought. In *The nature of thought*, P W Jusczyk and R M Klein, editors. Lawrence Erlbaum Associates, Hillsdale, New Jersey, 1980, pp. 19–35.

W Heron. The pathology of boredom. *Sci. Am.*, January 1957.

J J Hopfield. Neural networks and physical systems with emergent collective computational abilities. *Proc. Natl. Acad. Sci. USA*, 79:2554–2558, 1982.

D H Hubel and T N Wiesel. Receptive fields, binocular interaction and functional architecture in the cat's visual cortex. *J. Physiol.*, 160:106–154,1962.

J Joyce. *Ulysses*. Modern Library, New York, 1934.

R M Klein. D.O. Hebb: An appreciation. In *The nature of thought*, P W Jusczyk and R M Klein, editors. Lawrence Erlbaum Associates, Hillsdale, New Jersey, 1980, pp. 1–18.

C R Legéndy. On the scheme by which the human brain stores information. *Math. Biosci.*, 1:555–597, 1967.

C R Legéndy. Three principles of brain function and structure. *Int. J. Neurosci.*, 6:237–254, 1975.

C A Lindbergh. *The Saturday Evening Post*, June 6, 1953.

W S McCulloch. Modes of functional organization of the cerebral cortex. *Fed. Proc.*, 6:448–452, 1947.

W S McCulloch. *Embodiments of mind*. MIT Press, Cambridge, Massachusetts 1965.

W S McCulloch and W H Pitts. A logical calculus of the ideas immanent in nervous activity. *Bull. Math. Biophys.*, 5:115–133, 1943.

R Melzak and T H Scott. The effects of early experience on the response to pain. *J. Comp. Physiol. Psychol.*, 50:155–161, 1957.

P M Milner. The cell assembly: Mark II. *Psychol. Rev.*, 64:242–252, 1957.

P M Milner. The mind and Donald O. Hebb. *Sci. Am.*, January 1993: 124–129.

M Minsky and S Papert. *Perceptrons: An introduction to computational geometry*. MIT Press, Cambridge, Massachusetts 1969.

L A Necker. Observations on some remarkable phenomena seen in Switzerland; and an optical phenomenon which occurs on viewing of a crystal or geometrical solid. *Philos. Mag.*, Ser. 3, I:329–337, 1832.

N J Nilsson. *Learning machines*. McGraw-Hill, New York, 1965.

G Palm. Toward a theory of cell assemblies. *Biol. Cybern.*, 39:181–194, 1981.

G Palm. Cell assemblies, coherence, and corticohippocampal interplay. *Hippocampus*, 3:219–226, 1993.

W Penfield and P Perot. The brain's record of auditory and visual experience—a final summary and discussion. *Brain*, 86:595–696, 1963.

W Pitts and W S McCulloch. How we know universals: The perception of auditory and visual form. *Bull. Math. Biophys.*, 9:127–147, 1947.

R M Pritchard. Stabilized images on the retina. *Sci. Am.*, 204:72–79, June 1961.

R M Pritchard, W Heron, and D O Hebb. Visual perception approached by the method of stabilized images. *Can. J. Psychol.*, 14:67–77,1960.

M S Rabinovitch and H E Rosvold. A closed field intelligence test for rats. *Can. J. Psychol.*, 4:122–128, 1951.

A Rapoport. "Ignition" phenomena in random nets. *Bull. Math. Biophys.*, 14:35–44, 1952.

P Rakic. Local circuit neurons. *Neurosci. Res. Prog. Bull.*, 13, 1975.

A H Riesen. The development of visual perception in man and chimpanzee. *Science*, 106:107–108, 1947.

N Rochester, J H Holland, L H Haibt, and W L Duda. Tests on a cell assembly theory of the action of a brain using a large digital computer. *Trans. IRE Information Theory*, 2:80–93, 1956.

F Rosenblatt. The Perceptron: A probabilistic model for information storage and organization in the brain. *Psych. Rev.*, 65:386–408, 1958.

H Schröder. Über eine optische Inversion. *Annal. Physik Chemie*, 181:298–311, 1858.

A C Scott. Neurodynamics (a critical survey). *J. Math. Psychol.*, 15:1-45, 1977.

M von Senden. *Space and sight: The perception of space and shape in the congenitally blind before and after operation.* Methuen & Co., London, 1960. A republication of *Raum- und Gestaltauffassung bei Operierten Blindgeborenen vor und nach der Operation.* Barth, Leipzig, 1932.

C S Sherrington. *Man on his nature*, second edition. Cambridge University Press, Cambridge, 1951.

D R Smith and C H Davidson. Maintained activity in neural nets. *J. Assoc. Comp. Mach.*, 9:268–279, 1962.

W R Thompson and W Heron. The effects of restricting experience on the problem-solving capacity of dogs. *Can. J. Psychol.*, 8:17–31, 1954.

B L Wellman. Iowa studies on the effect of schooling. *Yearbook of the National Society for the Study of Education*, 39:377–399, 1940.

H White. The formation of cell assemblies. *Bull. Math. Biophys.*, 23:43–53, 1961.

B Widrow and J B Angell. Reliable, trainable networks for computing and control. *Aerospace Eng.*, September 1962:78–123.

T N Wiesel and D H Hubel. Effects of visual deprivation on morphology and physiology of cells in the cat's lateral geniculate body. *J. Neurophysiol.*, 26:978–1017, 1963.

M A Wilson and B L McNaughton. Dynamics of the hippocampal ensemble code for space. *Science*, 261:1055–1058, 1993.

A Century of
Brain-Mind Theory

*Either engineers must become poets
or poets must become engineers*

Norbert Wiener

C herished by poets, savored by gourmets, revered by lovers, and worshiped by mystics, consciousness is also a landscape of guilt and agony, remorse and pain. For the scientist and the philosopher, consciousness is intriguing because it continues—after centuries of study—to challenge our need to understand and to explain. We continue to wonder just what it *is*.

There is yet no consensus in the scientific community concerning the definition of consciousness, let alone the nature of the phenomenon. So instead of offering an explanation of consciousness, this chapter is an hors d'oeuvre tray, an eclectic survey of research on the mind-brain problem, together with many quotations. From this tray, the reader can sample the wide spectrum of theories that have been presented over the past century. There are many disagreements among these thinkers and writers, and some are rather strongly expressed. In general, my stance is to hold the tray— offering a *bruschetta* here and some *paté* there—with appropriate comments. But there are exceptions to this rule. When it is necessary to warn the reader that the *escargots* are off or the *vongole* doubtful, I shall do so. Please do not skip over the quotations, looking for the meat. All of these authors should be carefully read by any serious student of consciousness, so this selection has been designed to sharpen your appetite.

Finally, it should be emphasized that this survey is limited to studies of the scientific and philosophical bases of consciousness. The psychic dimensions and nature of consciousness have been of intense interest to spiritualists and mystics for several thousand years, but this activity—though clearly of great importance—lies beyond the bounds of the present discussion. A recent paper by Robert Forman (1994) provides a helpful introduction to studies of the many relationships between consciousness research and the mystic traditions.

WILLIAM JAMES

We begin our survey with some morsels from *The principles of psychology* by William James (brother of novelist Henry James), first published just over a century ago (1890). Two chapters of this celebrated work are of immediate interest in connection with the mind-brain problem, and the first of these is on "The automaton-theory." Here James considers the notion that the brain is merely a complex machine of many reflexes, each being modeled on the familiar spinal cord reflex. From this perspective, mind is reduced to the status of an "inert spectator" or an "epiphenomenon." To appreciate the consequences he argues that one should apply it to the most complicated examples.

> If we knew thoroughly the nervous system of Shakespeare, and as thoroughly all his environing conditions, we should be able to show why at a certain period of his life his hand came to trace on certain sheets of paper those crabbed little black marks which we for shortness' sake call the manuscript of Hamlet. We should understand the rationale of every erasure and alteration therein, and we should understand all this without in the slightest degree acknowledging the existence of the thoughts in Shakespeare's mind. The words and sentences would be taken, not as signs of anything beyond themselves, but as little outward facts, pure and simple.

This is the position that the reductive materialist undertakes to defend, and we shall return to it in future discussions. Is this *all* that can be said of Shakespeare? No, according to James, because

> nothing in all this could prevent us from giving an equally complete account of ... Shakespeare's spiritual history, an account in which every gleam of thought and emotion should find its place. The mind history would run alongside of the body history ... and each point in the one would correspond to, but not react upon, a point in the other. So the melody floats from the harp-string, but neither checks nor quickens its vibrations; so the shadow runs alongside the pedestrian, but in no way influences his steps.

Here James puts his finger squarely on an issue that continues to confound consciousness research to the present day. Any mechanistic explanation of Shakespeare's behavior, any reductive theory that explains his writing of *Hamlet* doesn't *need* an account of his thoughts and emotions. It is an awkward situation for the reductive materialist.

What proof is there for the automaton-theory? Some advance a continuity argument (that starts with the frog's spinal reflex and works upward) to support the reductive point of view, but such arguments work both ways. Thus one could as well begin with a fully developed consciousness and work down. James also rejects the tendency of laboratory scientists to avoid unmeasurable feelings as ignoring the essential problem. Logically one cannot do this and then be dogmatic about material causation, and he concludes that "to urge the automaton-theory upon us, as it is now urged, on purely *a priori* and *quasi*-metaphysical grounds, is "an *unwarrantable impertinence in the present state of psychology.*" Also consciousness should continue to be taken seriously, in James's opinion, because it is useful for survival, and since it is useful, "automaton-theory must succumb to the theory of common sense."

Having dispatched the automaton-theory, James turns to "The mind-stuff theory." Here—I must point out—he displays a confusion about the natures of *linear* and *nonlinear* interactions that was not unusual among scientists at the turn of the century. Arguing from this confused perspective, he finds the theory of mental units "compounding with themselves" or "integrating" (as in Figure 23) to be logically unintelligible:

> In other words, no possible number of entities (call them as you like, whether forces, material particles, or mental elements) can sum *themselves* together. Each remains, in the sum, what it always was; and the sum itself exists only *for a bystander* who happens to overlook the units and to apprehend the sum as such; or else it exists in the shape of some other *effect* on an entity external to the sum itself. Let it not be objected that H_2 and O combine of themselves into 'water' and thenceforward exhibit new properties. They do not. The water is just the old atoms in the new position, H–O–H; the 'new properties' are just their combined *effects*, when in this position, upon external media, such as our sense-organs and the various reagents on which water may exert its properties and be known.

I emphasize this point because James, as the preeminent American philosopher of his day, has had enormous influence on our thought, and this influence continues to the present. But this view is at variance with the general thesis of this book and, in particular, our understanding of the nature of water. Its wetness and its high dielectric constant—as we saw in Chapter 2—are *emergent* properties that water does *not* share with hydrogen and oxygen. It seems that James had forgotten his chemistry. Further on he asserts:

> The associationists say the mind is constituted by a multiplicity of distinct 'ideas' *associated* into a unity. There is, they say, an idea of a, and also an idea of b. *Therefore*, they say, there is an idea of $a + b$, or of a and b together. Which is like saying that the mathematical square of a plus that of b is equal to the square of $a + b$, a palpable untruth. Idea of $a +$ idea of b is *not* identical with idea of $(a + b)$. It is one, they are two; in it, what knows a also knows b; in them what knows a is expressly posited as not knowing b etc. In short, the two separate ideas can never by any logic be made to figure as one and the same thing as the 'associated' idea.

Again, I disagree. An identical problem is discussed in detail in Appendix A from the perspective of modern nonlinear theory. If the escape energy of child a on a trampoline is a^2, and that of child b is b^2, then the escape energy of both children together is $(a + b)^2$. It is indeed true that $(a^2 + b^2)$ is not equal to $(a + b)^2$. Instead it is equal to $(a^2 + 2ab + b^2)$, but that is exactly the point. The change in escape energy of $2ab$ when the two children are on the trampoline together is just the binding energy that holds them together. Although the associationists had it right, James had hereby launched the specious *binding problem*, which continues to bedevil research in cognitive neuroscience. From this perspective, he proceeds to an incorrect conclusion that I quote because it remains in circulation after a century. He claims that his argument

> holds good against any talk about self-compounding amongst feelings, against any "blending," or "complication," or "mental chemistry," or "psychic synthesis," which supposes a resultant consciousness to float off from the constituent *per se*, in the absence of a supernumerary principle of consciousness which they may affect. The mind-stuff theory, in short, is unintelligible. Atoms of feeling cannot compose higher feelings, any more than atoms of matter can compose physical things! The "things," for a clear-headed atomistic evolutionist, are not. Nothing is but the everlasting atoms.

Thus was James trapped between the Scylla of reductive materialism and the Charybdis of dualism. Being too much of a realist to ignore consciousness, he cautiously chose dualism in the following terms:

> I confess, therefore, that to posit a soul influenced in some mysterious way by its brain-states and responding to them by conscious affections of its own, seems to me the line of least logical resistance, so far as we have yet attained.

GEORGE SANTAYANA

An important attribute of consciousness, as it is ordinarily experienced, is the aesthetic sense, the ability to perceive beauty. Just a century ago, George Santayana, a young colleague and former student of James, published a small book on this topic, *The sense of beauty* (1896). Based on lectures delivered at Harvard College and written from the perspective of natural philosophy, this book has established itself as a classic of American literature.

While recognizing that both "ethics and æsthetics have suffered much from the prejudice against the subjective," Santayana approaches his subject from a psychological perspective as a phenomenon of mind and a product of mental evolution. He seeks to know "how an ideal is formed in the mind."

From the viewpoint of philosophy, aesthetics deals with a theory of values, which are "ultimately irrational," and his first consideration is to establish "the meaning and conditions of value." Thus:

Since the days of Descartes it has been a conception familiar to philosophers that every visible event in nature might be explained by previous visible events, and that all motion, for instance, of the tongue in speech, or of the hand in painting, might have merely physical causes. If consciousness is thus accessory to life and not essential to it, the race of man might have existed upon the earth and acquired all the arts necessary for its subsistence without possessing a single sensation, idea, or emotion. Natural selection might have secured the survival of those automata which made useful reactions upon their environment. An instinct of self-preservation would have been developed, dangers would have been shunned without being feared, and injuries revenged without being felt.

This, of course, is just the point that was made by William James in relation to the activities of Shakespeare. It was clearly in the air of philosophical discourse a century ago. Santayana continues:

In such a world there might have come to be the most perfect organization. There would have been what we should call the expression of the deepest interests and the apparent pursuit of conceived goods. For there would have been spontaneous and ingrained tendencies to avoid certain contingencies and to produce others; all the dumb show and evidence of thinking would have been patent to the observer. Yet there would surely have been no thinking, no expectation, and no conscious achievement in the whole process.

The onlooker might have feigned ends and objects of forethought, as we do in the case of water that seeks its own level, or in that of the vacuum which nature abhors. But the particles of matter would have remained unconscious of their collocation, and all nature would have been insensible of their changing arrangement. We only, the possible spectators of that process, by virtue of our own interests and habits, could see any progress or culmination in it. We should see culmination where the results obtained satisfied our practical aesthetic demands, and progress wherever such a satisfaction was approached. But apart from ourselves, and our human bias, we can see in such a mechanical world no element of value whatsoever. In removing consciousness, we have removed the possibility of worth.

These words, penned at the dawn of the twentieth century, are entirely relevant to the fervid discussions of robotics with which it is drawing to a close. The question of value (or the lack of it) is not restricted to the relationship between humans and machines. One wonders if Santayana foresaw some of the horrors coming over the human horizon as he wrote:

But it is not only in the absence of all consciousness that value would be removed from the world; by a less violent abstraction from the totality of human experience, we might conceive beings of a purely intellectual cast, minds in which the transformations of nature were mirrored without any emotion. Every event would then be noted, its relations would be observed, its recurrence might even be expected; but all this would happen without a shadow of desire, of pleasure, or of regret. No event would be repulsive, no situation terrible. We might, in a word, have a world of idea without a world of will. In this case, as completely as if consciousness were absent altogether, all value and excellence would be gone. So that for the existence of good in any form it is not merely consciousness but emotional consciousness that is needed. Observation will not do, appreciation is required.

Consciousness, according to Santayana, is useful not only for the survival of a species but also for the development of human culture.

BEHAVIORISM

It is difficult to verify the existence of consciousness because it is, somehow, internal. I can't experience your feelings and you can't experience mine. The subject is uncomfortable for one who wishes to deal only with data that can be objectively measured. What can one do? How should one deal with the elusive gap between a mechanistic and a dualistic picture of the mind?

One way to deal with the problem is to ignore it. This was the position of the behaviorists as expounded by John B. Watson in 1924. (He is not to be confused with James D. Watson, who discovered the double helix structure of DNA together with Francis Crick.) Put off perhaps by the animistic overtones in James's talk of the soul, behaviorists took up the automaton-theory in order to place psychology on a scientific basis. One of the most able defenses of "the experimental analysis of behavior" was put forward by Burrhus F. Skinner in *Beyond freedom and dignity* (1972). Written in the declining years of his considerable influence, this book went into nine hardback printings as a reviewer for the *New York Times* enthused, "If you plan to read only one book this year, this is probably the one you should choose!"

In this book Skinner distances himself from "methodological behaviorism," which limits itself to considering only what can be observed in public. The scientific analysis of behavior is often accused of ignoring consciousness, he admits, and he deems this a serious charge. It is allowed that mental processes "may exist, but they are ruled out of scientific consideration by their nature." All behavior would be unconscious without the aid of a verbal community, and more definitely he states that:

Consciousness is a social product.

Throughout the book Skinner is concerned with abolishing the concept of the *autonomous man*: "the inner man, the homunculus, the possessing demon, the man defended by the literatures of freedom and dignity." Instead of wasting time and effort trying to understand this fictitious inner person, he would rely on the methods of physics and biology to construct a scientific psychology that must necessarily represent human behavior as unintentional. It is not merely a question of scientific truth; this approach is necessary in order to solve the problems of society. (These problems seemed particularly acute in the United States circa 1970.) Thus he asserted:

> What we need is a technology of behavior. We could solve our problems quickly enough if we could adjust the growth of the world's population as precisely as we adjust the course of a space ship, or improve agriculture and industry with some of the confidence with which we accelerate high-energy particles, or move toward a peaceful world with something like the steady progress with which physics has approached absolute zero (even though both remain presumably out of reach). But a behavioral technology comparable in power and precision to physical and biological technology is lacking, and those who do not find the very possibility ridiculous are more likely to be frightened by it than reassured. That is how far we are from "understanding human issues" in the sense in which physics and biology understand their fields, and how far we are from preventing the catastrophe toward which the world seems to be inexorably moving.

Thus is the behaviorist concept of the mind revealed as a response to the need to control the chaos in which we find ourselves. But the fundamental question, of course, is not what we need but what we can have.

Although Donald Hebb's work did much to undermine the tenets of the behaviorists, he had respect for their contributions to the development of psychological research. In a section entitled "Nonsense about behaviorism" from his *Essay on mind*, which appeared in 1980, he took to task those who

> regard behaviorism and John B. Watson as somehow beyond the pale, intellectually speaking, not fit topics for reasoned criticism. For them Watson was not only wrong, he was perversely wrong— a view that is itself perversely out of line with the facts and a prime example of that weakness in psychology and social science for adopting extreme positions in matters that call for a more discriminating judgement. Such a view of Watson would not survive if those who hold it would take the trouble to look into the record, instead of contenting themselves with second- and third-hand criticism.

Hebb recalls for us that Watson did two things. First, he formulated a general method for psychological research that avoids introspection and relies entirely on objective evidence. None can complain about this. An experimental scientist is free to gather data under any rules that seem appropriate. But Watson did more. He went on to propose a theory that was both "unpalatable" and "outrageous" and, unfortunately, his critics did not restrict themselves to reasoned response.

With respect to Skinner, Hebb generously concludes:

> He is a brilliant experimentalist with a record of valuable contributions to human welfare, a fact that one would not detect from the kind of comments made in the journals of social science. One may well disagree with his radical positivism, as I do, but this should not blind the critic to the fact of his outstanding contribution. It is astonishing that the proposal of *Beyond Freedom and Dignity* to substitute positive reward for punishment and make criminals not want to be criminals should have appeared to a wide range of social scientists as a vicious program of thought control. ... To talk of behaviorism as blind incompetence is ignorance or prejudice or both. It is certainly not the mark of scholarship.

DONALD HEBB

But what does Hebb have to say about the relationship between the brain and the mind? Individual learning, he pointed out in 1949, does not rise along a smooth curve in a real psychological experiment, but exhibits fluctuations from day to day. The simple textbook curves that show steady improvement with practice are obtained by averaging measurements on many subjects. Although this variability is explained in different ways by writers with diverse theoretical perspectives, Hebb took the view that much of it arises from changes in the attitudes of the subjects toward the learning tasks. Related to this effect are periods of directed behavior or of spontaneous play, lasting longer in the dog or cat than in the rat and longer still in humans, as one should expect from the ratios of association to sensory areas of the neocortex. "There can be no explanation of learning and problem-solving in any mammal," he insisted, "without reference to the persisting central neural influence that sustains activity in one particular direction" (Hebb, 1949).

A voluntary or conscious act is influenced by both sensory and central facilitations, and the muscular reaction shows "motor equivalence" or the tendency to reach a constant end effect under the buffets of changing conditions. In the heyday of behaviorism, he wrote (1949):

> Empirically, in behavior, the most important mark of consciousness is the continually changing selective responsiveness to different aspects of a familiar environment, the *unchanging* responsiveness to unusual or unexpected events, together with the continual presence of 'purpose,' 'means-end-readiness,' or motor equivalence.

In 1949 Hebb was not specific about the mechanism of consciousness, but his position is limned more carefully in 1980, when he wanted his use of the terms "mind" and "consciousness" to be understood. He viewed the mind as that aspect of the brain that controls its behavior and noted that such control might be either "spiritual or immaterial" in nature or a "physical activity of the brain." Viewing mind as spirit is vitalism, which has no place in the scientific theories of modern biology. Thus he rejects dualism as a "stumbling block for the scientific approach to understanding man." Although dualism cannot be disproved, the role of science is to proceed on the assumption that it is wrong and see how much progress can be made. Hebb put his general position as follows:

> Mind is the capacity for thought,
> consciousness is a present activity of thought,
> and thought itself is an activity of the brain.

ERWIN SCHRÖDINGER

The scientific activities of Erwin Schrödinger intersect all the levels of discourse in this book. Chapter 2 was devoted to his seminal contributions to the theory of atomic structure and also to his famous cat, which continues to be of interest to those who would use quantum theory to understand the nature of biological reality. In Chapter 3, we noted Schrödinger's seminal suggestion for the localization of the genetic code. In the published version of these 1943 lectures (Schrödinger, 1944) an epilogue was added with the title: "On determinism and free will." (Deemed offensive to the Roman Catholic Church, the introduction of this epilogue delayed publication of the book, although it ultimately sold over 100,000 copies.) In this epilogue he expressed his opinion that quantum indeterminacy is biologically irrelevant (see Appendix E) and the space-time events in a living being corresponding "to the activity of its mind, to its self-conscious or any other actions are... if not strictly deterministic at any rate statistico-deterministic." (As in public opinion polls, the per cent error is $\pm 100/\sqrt{n}$, where n is the number of questionnaires or the number of particles.)

Then he took up the central issue of this chapter by stating two premises. The first is: "My body functions as a pure mechanism according to the Laws of Nature." This is the Scylla of James's automaton theory. The Charybdis of mental reality appeared in his second premise: "Yet I know, by incontrovertible direct experience, that I am directing its motions, of which I foresee the effects, that may be fateful and all-important, in which case I feel and take full responsibility for them." From these two premises, Schrödinger felt forced to infer that "I am the person, if any, who controls the 'motion of the atoms' according to the Laws of Nature."

In 1956 Schrödinger was invited to present the Tarner Lectures at Trinity College, Cambridge, and the subject he chose was "Mind and matter."

Although ill health prevented him from delivering them, these lectures were read in his stead from his notes and published in a short book (Schrödinger, 1958) that serves as a sequel to his earlier work. The first chapter of *Mind and matter* is entitled "The physical basis of consciousness" and in it he immediately posed the question:

What kind of material process is directly associated with consciousness?

Rejecting the reductionist view that consciousness is an outgrowth of nervous activity as intellectually "bankrupt," he pointed out that many neural processes are not accompanied by consciousness and concluded that "consciousness is the tutor who supervises the education of the living substance, but leaves his pupil alone for all those tasks for which he is already sufficiently trained." Or in different words:

Consciousness is associated with the learning of the living substance; its *knowing how* is unconscious.

Thus evolution involves a form of consciousness, and he used this perspective to sketch a scientific basis for a theory of ethics, although he protested, "I am a scientist, not a teacher of morals." This theory can be described as a struggle between the egoism of a solitary animal and the unselfish sharing of a social animal. We shall return to this question in Chapter 8.

Clearly influenced by Sherrington, Schrödinger discussed the "principle of objectivation" or the hypothesis of the real world around us. In order to view this world objectively, he argued, we step back "into the part of an onlooker who does not belong to the world," but the logic of this is hidden from us by two circumstances:

> First my own body (to which my own mental activity is so very directly and intimately linked) forms part of the object (the real world around me) that I construct out of my sensations, perceptions and memories. Secondly, the bodies of other people form part of this objective world. Now I have very good reasons for believing that these other bodies are also linked up with, or are, as it were, the seats of spheres of consciousness..., yet I have absolutely no direct access to any of them. Hence I am inclined to take them as something objective, as forming part of the real world around me. I so to speak put my own sentient self (which had constructed this world as a mental product) back into it— with the pandemonium of disastrous logical consequences that flow from the aforesaid chain of faulty conclusions.

The first of two "most blatant" contradictions that he noted was our surprise at finding our world picture "colorless, cold, and mute." We should not wonder that perceptions we all know—the color of a rose in the light of late afternoon, the warmth of an open fire after a day in the mountains, the sound of a violin as it strives upward to the stars—are missing in a world model from which we have removed ourselves. A second contradiction was seen in our fruitless search for the locus of the mind. Since our objective world has been constructed by removing our mind from it, we should not be surprised not to find it there.

Schrödinger saw the habit of localizing consciousness inside the body as merely symbolic. Looking inside the head one sees "tens of thousands of contacts being opened and blocked within every split second," but nowhere, "however far physiology advances" will one find the feelings that constitute reality for us all. As we saw in the previous chapter, this is exactly the question raised by Charles Sherrington with respect to seeing a star. Again yawns the gap between between reductive materialism and psychic reality.

The reason, according to Schrödinger, that we cannot find our sentient self in the world picture of science can be simply stated:

It is itself that world picture.

CAN A MACHINE THINK?

In 1950 Alan Turing published a paper with this provocative title, but he chose not to answer it. Instead he asked a related question: "Can a machine pass for a human in an imitation game?"

In this game the *subject* (either the machine being tested or a human) is hidden from an *interrogator* with whom he or she (or it) communicates via printed messages. The interrogator tries to determine whether or not the subject is human, and if a machine does about as well as a human, it is judged to have passed the test.

This *Turing test*, as it is now called, provides an operational definition of the verb *to think*, but there is further difficulty in defining the term *machine*. Although one would like to permit any sort of fabrication technique, even those that cannot be fully described, it is necessary to exclude the offspring of a male and a female computer engineer or an individual raised from a cloned human cell. To avoid such difficulties, Turing required the machine to be a digital computer. Thus the original question is reduced to the more limited and precise question:

Can a digital computer pass for human in the imitation game?

Turing stated his belief "that in about fifty years' time it will be possible to programme computers with a storage capacity of about 10^9, and to make them play the imitation game so well that an average interrogator will not have more than 70 per cent chance of making the right identification after five minutes of questioning."

Turing considered several objections to his belief. These included the idea that thinking is done by a God-given soul and an objection based on a horror of thinking machines, both of which he rejected out of hand. Some consideration was given to mathematical objections based on Gödel's theorem (Kurt Gödel showed in 1931 that any nontrivial logic system— arithmetic, for example—will have true statements that cannot be proven with a given set of axioms), to biological objections that machines lack consciousness or that the nervous system is not a digital machine, and to objections related to observations of extrasensory perception, all of which he found wanting in one way or another.

A more serious objection, in Turing's opinion, is that a machine does only what it is told to do; it does not do anything new. Thinking, on the other hand, is an act of creation, which leads one back to the question of consciousness. Those who advance this objection, Turing believed, would probably be willing to accept the results of an imitation test.

We now have the opportunity of reviewing Turing's predictions after a period of "about fifty years' time." In one respect he was clearly correct: the hard disk memory capacity of a standard lap-top computer is currently 10^9 bits or more, which is enough to store the *Encyclopædia Britannica*. Although—as I secretly suspect—some members of the virtual community of cyberspace may be computers trained to hitchhike on the electronic highway, his predictions of computer intelligence seem overly optimistic. There is as yet no good computer code for translating from one natural language to another.

ARE WE MACHINES?

In a short but thoughtful paper that appeared in the *Proceedings of the American Philosophical Society* in 1969, Eugene Wigner, a distinguished mathematical physicist, asked the question "Are we machines?" In this paper he discussed two standard views on the nature of life. The first is mechanistic. Life is a process that can be explained on the basis of the "ordinary laws of physics and chemistry." Wigner noted that this was the view of most physical scientists until about 1940, and it was still popular among biologists. The second view is nonmechanistic. "In addition to inanimate objects, there is also a mind, a consciousness, and . . . a description of the motion of the atoms and the molecules of living beings cannot give a complete description of their life."

From the perspective of a physicist, Wigner pointed out, the ordinary laws of physics and chemistry have been replaced by the laws of quantum mechanics. As we saw in Chapter 2, the positions and velocities of atoms cannot be exactly fixed, so the new theory is couched in terms of "probabilities of the outcomes of measurements, given the character and outcome of the preceding measurement" rather than as precise initial data. Although the concept of the position of the atoms is often useful,

> the primitive concept according to quantum mechanical theory is not this position but the knowledge of the observer concerning the state of the molecule, and it seems unreasonable to describe the basic concept—the content of the consciousness of the observer—in terms of a derived one, valid only under certain conditions, as is the concept of the positions of atoms.

As a simple example of an inadequate or incomplete theory, Wigner proposed the wave equation for electric field intensity in a vacuum—see Appendix I—without knowledge of the magnetic field intensity. He claimed that the mechanical model of life is similarly incomplete because only a partial description of reality is presently available to us. "There can be very little interest in the complicated chemical reactions that go on in my brain," he states,

> and in minute variations of these reactions, as long as the state of my mind, the pleasure I experience or the pain I feel, cannot be inferred from the knowledge of these reactions. It is not possible to prove on the basis of present physical theories that such an inference is impossible; it is true that we have at present not even the vaguest idea how to connect the physico-chemical processes with the state of the mind.

Since Wigner wrote, there has been some progress in the field of psychophar-macology. Knowledge accumulates on the effects of alcohol, Prozac, cocaine serotonin, caffeine, etc. and where in the brain they act, but we still do not know *how* they act.

It may be that the present laws of physics are sufficient to understand the nature of mind. In this case it would be required to find an equation that could relate consciousness to the underlying physical world. Wigner, however, believed that it is necessary to *change* the laws of physics and not merely to reinterpret them. On the basis of past experience in science, it is to be expected that such changes would introduce new concepts of presently unappreciated subtlety and abstraction.

Are we machines? Wigner replies:

> If predictability is admitted as
> a necessary condition for being a machine,
> we are not machines.

FOREBRAIN COMMISSUROTOMY

The human brain is segmented into two hemispheres, which are connected through a massive bundle of several hundred million fibers called the corpus callosum (see Figure 20). In the 1940s various clinical reports indicated that complete surgical section of the corpus callosum (a procedure known as forebrain commissurotomy) did not result in obvious psychological defects. Stimulated by such observations, Roger Sperry and his colleagues began a series of psychological studies on primates and humans, the results of which over the next quarter century are reviewed and evaluated in Sperry (1977). (In human patients, of course, the surgery was performed to control intractable epilepsy.)

Subjects were placed in experimental environments in which their left hemispheres received images that differed from those received by the right hemispheres: typically split photographs of faces for humans and crosses or circles for primates. Surprisingly, these studies indicated that each surgically disconnected hemisphere was able to "sense, perceive, learn, and remember independently of the other." Since the temporal lobe of the left hemisphere is the locus for human speech, it was straightforward to discuss with this "vocal hemisphere" what it perceived, but a "disgusted shaking of the head or irked facial expressions" were often initiated by the minor (right) hemisphere when it judged its partner to be in error. Emotional tone, on the other hand, appeared to be rapidly transmitted from one hemisphere to the other, probably through the brain stem. On the basis of these experiments, Sperry concluded that both hemispheres are independently conscious following commissurotomy, although he noted that this point of view is disputed.

Dramatic further evidence for individual seats of consciousness in the two hemispheres is provided by recent observations of neurologist Victor Mark on a college-educated, 33-year-old woman who underwent complete surgical section of the corpus callosum (Mark, 1995). This woman had always used her left hand for writing but favored her right hand for other activities, and there was presurgical evidence that her speech could be organized by either hemisphere. (The evidence is from her response to the Wada test, in which a barbiturate is injected separately into each carotid artery to determine the hemisphere that is mainly involved with language.) Under certain circumstances this patient would contradict herself. "When answering questions about her epilepsy or performing challenging psychological tests..., she tended to alternate rapidly among the responses 'Yes,' 'No!,' and 'That's not right!' " She had difficulty accepting such behavior and often became upset or withdrew from the conflict.

At one session she noted that she had no feelings in her left hand. When Mark repeated her statement, "She said she was *not* numb, and then the torrent of alternating 'Yes!' and 'No!' replies ensued, followed by a despairing 'I don't know!' " Mark notes that the surgery probably rendered her left hemisphere unaware of the feelings from her left hand. Further indication of this appeared when she was asked to point to the words YES or NO in reply to the question "Is your left hand numb?" In response, "The left hand jabbed at the word 'NO' and the right hand pointed to 'YES'. The patient became emotionally upset by the lack of unanimity and furiously and repeatedly tried to indicate the 'correct' answer, but with the same results. Ultimately the left hand forced aside the right and *covered* the word 'YES'!"

Although commissurotomy patients do not ordinarily show conflict to this extent, this woman appeared to have independent neural configurations in the two hemispheres that "simultaneously interpreted stimuli, formulated opinions, and communicated unambiguously in a symbolic manner." Although the corpus callosum helps the brain to remain vigilant and attentive, Mark concluded from these observations that "the essential structure for consciousness is one intact hemisphere connected to the brain stem."

In considering the implications of such split-brain experiments for the nature of consciousness in the normal brain, Sperry draws conclusions that lie close to the main point of this book. Deeply influenced by the tenets of behaviorism and reductive materialism, he initially resisted the "unthinkable thought" that the conscious mind could exert causal influence on behavior, but eventually he saw consciousness to be

a dynamic emergent of brain activity, neither identical with, nor reducible to the neural events of which it is mainly composed. Further, consciousness was not conceived as an epiphenomenon, inner aspect, or other passive correlate of brain processing but rather to be an active integral part of the cerebral process itself, exerting potent causal effects in the interplay of cerebral operations. In a position of top command at the highest levels in the hierarchy of brain organization, the subjective properties were seen to exert control over the biophysical and chemical activities at subordinate levels.

He came to a view of brain in which consciousness was the "crowning achievement of evolution," although he emphasized that no direct proof for this perspective is available, as is also the case for the position of the behaviorist. Furthermore he insisted that:

The causal power attributed to subjective properties is nothing mystical. It is seen to reside in the hierarchical organization of the nervous system combined with the universal power of any whole over its parts. Any system that coheres as a whole, acting, reacting, and interacting as a unit, has systematic organizational properties of the system as a whole that determine its behavior as an entity, and control thereby at the same time the course and fate of its components. The whole has properties as a system that are not reducible to the properties of the parts, and the properties at higher levels exert causal control over those at lower levels. In the case of brain function, the conscious properties of high-order brain activity determine the course of neural events at lower levels.

In other words:

> The meaning of the message will not be found
> in the chemistry of the ink.

KARL POPPER AND JOHN ECCLES

Essential reading for all who are interested in the mind-body problem is a carefully and imaginatively prepared book entitled *The self and its brain* by Karl Popper and John Eccles (1977). Subtitled *An argument for interactionism*, this work deftly blends the philosophical insights of Popper with the scientific knowledge of Eccles, a celebrated electrophysiologist, through book-length sections written by each and followed by transcriptions of twelve of their conversations or dialogues. At the outset Popper establishes his central position:

**I wish to state clearly and unambiguously
that I am convinced that selves exist.**

Beyond this, his fundamental idea is that reality should be viewed in three different aspects:

World 1. This is the objective world of Schrödinger, the realm sketched in Chapters 2 through 6 of this book. In Popper's terms, it is the "universe of physical entities" in which the various levels of physical science interact with each other through their appropriate dynamic laws. A reductive materialist would hold that this includes all of reality.

World 2. Beyond the universe of physical entities, we experience—whether we choose to admit it or not—the psychic reality of an inner life that includes a vast and multicolored tapestry of desires, ideas, pains, joys, sorrows, loves, schemes, strivings, and songs that are jumbled together with memories of the past and hopes and fears for the future. This inner reality is called World 2, and of course there are many of them: one for each of us.

World 3. The world of human culture in the broadest sense of the term is called World 3, and it includes all "the products of the human mind, such as stories, explanatory myths, tools, scientific theories (whether true or false), scientific problems, social institutions, and works of art."

Popper notes that a similar perspective goes back to Plato who distin-guished between the world of "visible objects" (World 1), the world of "intelligible objects" (World 3), and the "states of the soul" (World 2).

Interactions between Worlds 1, 2, and 3 are rather involved as is indi-cated by the following considerations. World 3 items are often embodied in World 1 objects. Thus a novel might exist as a printed book (coded in the ink that forms the letters on the pages) or on a floppy disk or both, but the World 3 object has an existence that is independent of its World 1 manifestation. World 3 items may also exist without reference to World 1. Examples are the primitive oral traditions that were carried on before the discovery or introduction of written language, in which a story resided in the mind (World 2) of a village elder. World 3 items may exist, unembodied and waiting to be found, without reference to either World 1 or World 2. As an example, Popper cites the properties of the prime numbers before their dis-covery but after the invention or discovery of the natural numbers. World 2 items are real in the sense that their existence can influence events in World 1. Feelings of fear or love, for example, can make one act in unexpected ways with unintended effects in World 1. A barbarian prince's love for an exotic princess leads to a marriage that brings fresh taste (World 3) into the realm, and begins the construction of new World 1 objects. Unembodied World 3 objects are also real in the same sense. The possibility of making an atomic bomb, for example, led to its construction with World 1 consequences that are not yet fully appreciated. And so on. . .

Careful consideration of World 3, according to Popper, can illuminate the mind-body problem. He presents three arguments to support this view, and the first is this: Although World 3 objects are abstract (even more so than physical forces), they are also real, for they can change World 1. But World 3 affects World 1 only through human intervention, because it involves a World 2 process. Thus "We therefore have to admit that both World 3 objects and the processes of World 2 are real—even though we may not like this admission, out of deference, say, to the great tradition of materialism."

Popper's second argument is that an understanding of the relationship between World 3 and World 2 (a relationship we shall consider in the next chapter) may help us to understand the relationship between World 2 and World 1, and this relationship is part of the mind-body problem.

His third argument is related to the nature of human language. "The capacity to learn a language—and even a strong need to learn a language—is, it appears, part of the genetic make-up of man. By contrast, the actual learning of a particular language . . . is . . . a cultural process, a World 3 regulated process, . . . in which genetically based dispositions, evolved by natural selection, somewhat overlap with and interact with a conscious process of exploration and learning, based on cultural evolution. This supports the idea of an interaction between World 3 and World 1; and in view of our earlier arguments, it supports the existence of World 2."

The sense of self, in Popper's view, is based in all three worlds but especially in World 3, and it is important to recognize that we must *learn* to be selves. (Compare von Senden's descriptions of "learning to see" in the previous chapter.) A fully dimensioned human being does not evolve in a social vacuum; without social interactions one devolves into a brutish *Homo ferus*.

Popper doesn't like the question "What is consciousness?" because "What is. . . ?" questions often degenerate into verbalism, which he tries to avoid. Moreover, he doesn't claim to solve the problems related to consciousness, but rather to sketch a program of research that may deepen our understanding of the concept. There is nothing mystical about his position. "It is fairly clear," he states, "that the identity and integrity of the self have a physical basis. This seems to be centered in our brain." Further on he conjectures that "the flawless transplantation of a brain, were it possible, would amount to a transference of the mind, of the self. I think that most physicalists and non-physicalists would agree with this."

To the question "What are the biological achievements that are helped by consciousness?" Popper provides three answers. First is the solution of nonroutine problems. Writing and intelligent speech are good examples: one must constantly construct *new* sentences. (Perhaps computing machines need to be constructed in a way that includes consciousness before they can translate languages and pass the Turing test.) Second, consciousness is necessary in order to formulate new theoretical perspectives or expectations. It is the unexpected event that requires conscious attention. Finally, in cases when one is faced with two alternate means to achieve a purpose, consciousness is needed to make a decision.

For his part, Eccles grounds the discussion in the facts of electrophysiology. This is important because

> Far too little consideration to the neuronal machinery involved in the various manifestations of the self-conscious mind has been given in the past. Philosophers presenting physicalist theories of the brain-mind problem... should build their philosophies upon the best available scientific understanding of the brain. Unfortunately, they are content with crude and antiquated information that often leads them into espousing erroneous ideas. There is a general tendency to overplay the scientific knowledge of the brain, which, regretfully, also is done by many scientists and scientific writers.

Eccles did not assert that present levels of scientific knowledge will solve the mind-brain problem, but he did claim "that our present knowledge should discredit the formulations of untenable theories and that it will give new insights into such fundamental problems as conscious perception, voluntary action, and conscious memory."

He begins with a survey of macroscopic and microscopic features of the cerebral cortex, the vast, layered structure covering the brain that is composed of six layers and laterally divided into motor and sensory transmitting areas. The cortex can be further subdivided into modules (3 millimeters long and 0.1 to 0.5 millimeters across) that are perpendicular to the surface and contain about 10^4 neurons each. These modules are found to be tied together by interconnections that allow them to build up "power" (or neural firing activity) while inhibiting the activities of neighboring modules. Eccles regards each of these modules as a "power unit" that struggles with its neighbors for neural activity, but this struggle is always under control.

> It is in this continuous interaction that we have to think of the subtlety of the whole neuronal machine of the human cerebral cortex composed of perhaps of one to two million modules each with up to 10,000 component neurones. We can only dimly imagine what is happening in the human cortex or indeed in the cortices of the higher mammals, but it is at a level of complexity, of dynamic complexity, immeasurably greater than anything else that has ever been discovered in the universe or created in computer technology."

The number of modules is estimated by Eccles by dividing the number of neurons in the neocortex (2×10^{10}) by the number of neurons in a module (10^4). If one supposes that each cell participates in more than one module and employs the countings of Legéndy or Hopfield, as discussed in the previous chapter, the estimate of the number of modules increases by several orders of magnitude.

Eccles considers the lessons to be learned from lesions of the brain, and in particular the work of Sperry (1977), whose observations lead Eccles to the conclusion that, "Conscious experience is uniquely and exclusively related to the dominant hemisphere." This conclusion is closely related to his central chapter "The self-conscious mind and the brain" in which he hypothesizes that the "self-conscious mind is an independent entity." This entity, he suggests, actively surveys and supervises the dynamic modules (sometimes called "large assemblages" or "colonies") of the cortex, and he proposes that "the unity of conscious experience comes not from an ultimate synthesis in the neural machinery but in the integrating action of the self-conscious mind on what it reads out from the immense diversity of neural activities in the liaison brain."

This is a dualist-interactionist explanation that assumes the self-conscious mind to be "superimposed upon the neural machinery." At certain sites of the cortex (the liaison areas), two-way interactions between mind and brain take place. Thus there are continuous subjective interactions indicated by

$$\text{World } 1 \rightleftharpoons \text{World } 2$$

in addition to the cultural interactions

$$\text{World } 1 \rightleftharpoons \text{World } 2 \rightleftharpoons \text{World } 3 \text{ ,}$$

where

$$\text{World } 1 \leftarrow \text{World } 2$$

indicates voluntary action and

$$\text{World } 1 \rightarrow \text{World } 2$$

implies conscious perception.

Eccles claims that this hypothesis is stronger and more definitive than previous dualistic postulates, and in support he cites four items of evidence: First, the experiences of the self-conscious mind display a *unitary character* that involves concentration on one thing at a given time. Second, although there is a relationship between the experiences in the self-conscious mind and neural events in the brain, they are not identical. It is "inconceivable that all-or-nothing pulses could be directly involved in the liaison with the self-conscious mind." Third, temporal discrepancies are observed between neural events and mental experiences. (As was noted in Figure 17, pulse timings are elusive even at the level of a single neuron.) In particular, physiologist Benjamin Libet and his colleagues have presented evidence that suggests a "subjective antedating" or backward referral in time of conscious perceptions (Libet, 1973; Libet et al., 1979). Eccles tentatively accepts this work as evidence that the self-conscious mind follows different dynamics from those of the brain, but one should be aware that these experimental results are open to alternate interpretations (Honderich, 1984; Libet, 1985). Finally, we experience the actions of the self-conscious mind on events in the brain.

The sites for interactions between the self-conscious mind and the brain occur at some of the dynamic modules but not all at a particular time. "Only in such assemblages can there be reliability and effectiveness."

There seems to be some disagreement between Popper and Eccles concerning the fundamental nature of the self-conscious mind. Thus we compare Popper's statement, "It is fairly clear that the identity and integrity of the self have a physical basis" with that of Eccles: "I believe that the reductionist strategy will fail in the attempt to account for the higher levels of conscious performance in the human brain." This question is discussed in Dialogue X, where Eccles asked Popper to discuss the relationship between World 1 and World 2.

Popper replied that it was a difficult problem and his ideas were not mature, but he went on to consider whether the World 2 hypothesis was in conflict with fundamental laws of physics, and in particular, with the law of energy conservation. Citing the impact of selection pressures on mutations as a heuristic model, and noting the openness of Newtonian mechanics to the forces of electromagnetism, and the openness of electromechanics to nuclear and weak forces, he tentatively concluded that

> we have, I think, in any case to postulate the openness of World 1 to World 2, while the mere openness of World 1 to an unknown part of World 1 would not help in solving the great problem that World 3 plans and theories bring about changes in World 1.

In response, Eccles drew attention to the views of both Schrödinger and Wigner that life cannot be explained by the ordinary laws of physics and chemistry.

BLINDSIGHT

Since some assert that consciousness is but a passive artifact of what the brain does (as a "melody floats from the harp-string"), it is interesting to consider experiments that show awareness of mental activity being altered. One such phenomenon—called blindsight by Lawrence Weiskrantz (1986)—involves unexpected abilities of surgical patients and experimental animals with portions of the visual cortex removed. Perhaps the best-documented case is that of D.B., an English male born in 1940, whose severe headaches became intolerable in his late twenties. Surgical removal of a malformation of the optical lobe in 1973 relieved his painful symptoms and allowed him to resume normal life, albeit with expected blindness in the left visual fields of both eyes.

Chance observations of D.B.'s response to visual stimulation in his blind area led Weiskrantz and his colleagues to conduct a careful series of experiments between 1973 and 1985, which are described in detail in Weiskrantz (1986). These studies involved repeated observations of reaching for randomly located targets, detection of "presence" versus "absence" of visual stimulation, visual acuity, threshold for movement, discrimination of orientation, discrimination of form (X vs. O, T vs. 4, A, C, D, R, or S, curved vs. straight triangles, squares vs. rectangles, and squares vs. diamonds), detection of a slow rate of onset, and detection of the direction of contrast. All such experiments showed statistically significant responses from the blind area, although D.B. insisted that he was "only guessing" because he was unaware of the stimulations.

With stimulation in his right (good) field, D.B. would report "seeing" or "not seeing," whereas in the left (blind) field, he would be "guessing x" or "guessing non-x"; thus a positivist might treat both as behavioral responses, but Weiskrantz argues that the reports of awareness and non-awareness must be taken seriously. In this context, *blindsight* is defined as a "visual capacity in a field defect in the absence of acknowledged awareness."

The awareness of visual processing seems to be switched off in blindsight, providing evidence for the view that consciousness is a mechanism that operates to some degree independently of underlying neural processes. Recent studies suggest that blindsight patients may also be sensitive to color (Stoerig and Cowey, 1989).

ROGER PENROSE

In his best-selling book *The emperor's new mind* (1989), Roger Penrose rejects the claims made by functionalists in the artificial intelligence community that what the brain does can be reduced to an algorithm and duplicated "in principle" on a digital computer. To this end he takes the general reader on a fascinating tour that includes a survey of computational algorithms, the nature of nonlinear mathematics, complexity and computability theory (Turing machines and Church's lambda calculus), classical dynamics (including general relativity), quantum theory and Schrödinger's cat, cosmology, quantum gravity, Gödel's incompleteness theorem, and the structure of the human brain. For our purposes the most interesting chapters are the first, entitled "Can a computer have a mind?" and the last, called "Where lies the physics of the mind?"

The belief is widespread, he writes, that "everything is a digital computer," and his aim is "to show why, and perhaps how, this need *not* be the case." He does not find it expedient to attempt a precise definition of consciousness, but (with James, Sherrington, Hebb, Schrödinger, and Popper and Eccles, among others) he considers what selective advantages might be realized by a species that evolves toward consciousness.

To support his claim that consciousness is nonalgorithmic—and therefore inaccessible to an algorithmic computer—he provides several examples of mathematical inspiration and nonverbal thought. Mathematics—in Penrose's view—is not invented but discovered, and the essential elements of a mathematical idea are most effectively first communicated in a vague and descriptive manner rather than precisely. He imagines

> that whenever the mind perceives a mathematical idea, it makes contact with Plato's world of mathematical concepts. (Recall that according to the Platonic viewpoint, mathematical ideas have an existence of their own, and inhabit an ideal Platonic world that is accessible via the intellect only.) When one 'sees' a mathematical truth, one's consciousness breaks through into this world of ideas, and makes direct contact with it ('accessible via the intellect').

This "world of ideas" is Plato's "world of intelligible objects" and is closely related to Popper's World 3.

Since classical dynamics, in Penrose's opinion, is "in principle" computable while quantum theory is not, he suggests that quantum theory may play a role in consciousness. He notes that there is one example where a single quantum of light (or a few quanta) can trigger neural activity—the retinal protein rhodopsin—and from this proof of existence he speculates that other cells might be found with corresponding sensitivities. But, as he indicated in a recent interview (1994a), he was by no means certain about this suggestion:

> Of all the views I put forward in *The Emperor's New Mind*, I was least comfortable with the idea that nerve signals could really be treated quantum mechanically; it always seemed a little bit hard to believe, although I was, in a way, pinning my faith on it.

Quite recently, Penrose has discussed the question of quantum consciousness in greater detail in *Shadows of the mind: A search for the missing science of consciousness* (1994b), which may be regarded as a sequel to his previous work. This book is divided into two parts, the first of which restates his arguments that mental activity is not computable. To clarify matters, he outlines four philosophical positions that one might assume:

A. All thinking is computation; in particular, feelings of conscious awareness are evoked merely by the carrying out of appropriate computations.

B. Awareness is a feature of the brain's physical action; and whereas any physical action can be simulated computationally, computational simulation cannot by itself evoke awareness.

C. Appropriate physical action of the brain evokes awareness, but this physical action cannot even be properly simulated computationally.

D. Awareness cannot be explained by physical, computational, or any other scientific terms.

A is the position of strong artificial intelligence or *functionalism*, and *D* is the position of the mystic. Both are rejected by Penrose so the choice is between *B* and *C*. *B*, he suggests, is the view that would generally be regarded as "scientific common sense" because the simulation of a physical process is not the same as the actual process. ("A computer simulation of a hurricane, for example, is certainly no hurricane!") Nonetheless, C is the position that Penrose believes to be closest to the truth.

View C holds that not all physical actions can be simulated on a computer, and Penrose argues—as did Wigner—that such noncomputable physical laws may lie outside the present purview of physics.

In the second part of his book, Penrose turns to physics and biology in an attempt to discover the seat of noncomputable physical action. In a short section entitled "Consciousness: new physics or 'emergent' phenomenon?" he allows that the reason consciousness appears only in brains "has something to do with the subtle and complex organization of the brain" but asserts that this is not a sufficient explanation. The organization of the brain must employ noncomputable physical effects.

> This picture differs markedly from a more commonly expressed view of the nature of consciousness (basically that of A) according to which consciousness awareness would be some sort of 'emergent phenomenon', arising merely as a feature of sufficient complexity or sophistication of action, and would not require any specific, new, underlying physical process, fundamentally different from those that we are already familiar with in the behaviour of inanimate matter. The case presented [here] argues differently, and it requires that there must be some subtle organization of the brain which is specifically tuned to take advantage of the suggested non-computable physics.

What is this noncomputable physics? Where does one find it? Not at the level of the neuron, for nerve impulses are macroscopic objects that are not restricted by quantum effects. But a one-celled animal, such as a paramecium, uses hairlike projections (*cilia*) to negotiate the obstacles of its little life without a nervous system, and the microtubules of the cytoskeleton, which were introduced in Chapter 3, seem to be implicated in this remarkable behavior. Since the axons, dendrites, and synapses of nerve cells are amply endowed with cytoskeletal structures (see Figure 8), Penrose turns to microtubules as a locus for the noncomputable physics of the mind.

> Let us then accept the possibility that the totality of microtubules in the cytoskeletons of a large family of the neurons in our brain may well take part in global quantum coherence—or at least that there is sufficient quantum entanglement between the states of different microtubules across the brain—so that an overall *classical* description of the collective actions of these microtubules is *not* appropriate.
>
> . . .
>
> On the view that I am tentatively putting forward, consciousness would be some manifestation of this quantum-entangled internal

cytoskeletal state and its involvement in the interplay between quantum and classical levels of activity. The computer-like classically interconnected system of neurons would be continually influenced by this cytoskeletal activity, as the manifestation of whatever it is that we refer to as 'free will'. The role of neurons, in this picture, is perhaps more like a *magnifying device* in which the smaller-scale cytoskeletal action is transferred to something which can influence other organs of the body—such as muscles. Accordingly, the neuron level of description that provides the currently fashionable picture of the brain and mind is a mere *shadow* of the deeper level of cytoskeletal action—and it is at this deeper level that we must seek the physical basis of *mind*!

While agreeing with Penrose that view C is the best choice of a philosophical perspective, I would search elsewhere for physical actions that cannot be simulated on a computer. He has given the concept of *emergence*—upon which the arguments of the present book are based—such short shrift that the term is not even listed in his index! While he searches for the mind at the lowest level of neural organization (the cytoskeletons that are inside the axons, dendrites, and synapses), I would seek it among the highest levels, among the interactions of complex cell assemblies, their phase sequences, and the vagaries of human culture. How does such a stark difference arise?

One answer to this question might be that I am unable—because of my intellectual limitations, interests in other matters, general laziness, or some combination of all three—to follow Penrose's arguments through their many details. Perhaps he is right, and I just don't get it.

Although this may be true, I think there is a deeper reason for our disagreement. In the world of science, the belief in reductive materialism is very strong. It is so strong that many—if not most—scientists don't even consider it to be a belief; it is seen as an evident fact of nature. But is it evident? Isn't a basic difficulty with the scientific description of consciousness that this deterministic belief does not square with everyday observations? Why is it so difficult for scientists—who have overthrown many myths of the past—to think the "unthinkable thought," to suppose that there are aspects of the natural world that cannot be reduced to laws of physics?

To assume the regularities that emerge at one level of description to be unrelated to those at lower levels does not require one to give up on science. As Erwin Schrödinger pointed out in the quotation at the end of Chapter 3, this is not the same as choosing view D from Roger Penrose's philosophical menu. We shall see in the next chapter that Franz Boas was no less a scientist for questioning the reductive materialism of his university days.

FRANCIS CRICK AND CHRISTOF KOCH

After making one's way through these philosophical discussions, it is refreshing to read a short but informative article by Francis Crick and Christof Koch entitled "Toward a neurobiological theory of consciousness" (1990). Considering that the subject is so important and so poorly understood, these authors found it remarkable that most of the work in the neurosciences does not mention consciousness or awareness, and they suggested that "the time is now ripe for an attack on the neural basis of consciousness." This paper is valuable because it sets aside several questions that are regarded as premature and concentrates on developing a strategy for obtaining useful experimental results. Their general approach is based on two assumptions: First, that there is something to be investigated, and second, that all the different aspects of consciousness involve a common mechanism. If we can understand the mechanism for one form of consciousness, it is tentatively assumed, we may understand the mechanism for them all. The reason behind this second assumption is that consciousness is such a peculiar phenomenon (like the murders in the *rue Morgue*) that only a particular explanation will work.

The topics to be set aside are these: First, since everyone has a rough idea of what consciousness is but no one understands it, it is premature to attempt a precise definition, and arguments about the function of consciousness are also premature for the same reason. Second, it is of little value to argue about whether animals below the higher mammals share consciousness, and since self-consciousness is assumed to be but a special case of ordinary consciousness, consideration of it is also premature. Third, since no neural theory will explain everything about consciousness, it is best to begin with a rough theory that might explain some aspects. Finally, questions of *qualia* (or whether your sense of the color red is the same as mine) should wait until we understand how we see color at all.

Having decided what *not* to consider, these authors suggest several questions that can be addressed with profit: When is an animal conscious? Where in the brain are the neural correlates of consciousness? What is the general character of neural behaviors associated, or not, with consciousness? Where are these neurons in the brain? Are they of any particular neuronal type? What is special (if anything) about their connections? What is special (if anything) about the way they are firing?

At this point Crick and Koch made an admittedly rather arbitrary decision to concentrate on the mammalian visual system. This limitation has the advantage of not requiring language, so nonhuman subjects can be studied, and it permits inputs from split-brain (Sperry, 1977) and blindsight (Weizkrantz, 1986) experiments.

Getting down to cases, they note that there are an "almost infinite" number of visual patterns (I would say an immense number) that can be recognized, so it is not possible to assume a "grandmother cell" that corresponds to each pattern. Thus recognition must be related to the activity of a *set* of neurons, but this leads immediately to the *binding problem* because recognition of a single pattern must involve neurons in several different visual areas. In order to be bound together, these authors argued, the participating neurons must carry a common label, and they suggest that electrochemical oscillations in the 40–70 hertz range bind the relevant neurons in short-term memory for a few seconds, an idea, they observe, that goes back to William James (1890). It is interesting to recall, from the previous chapter, that this frequency range coincides with that found effective by Wilder Penfield for the electrical ignition of experiential responses. In their words:

> We suggest that one of the functions of consciousness is to present the result of various underlying computations and that this involves an attentional mechanism that temporarily binds the neurons together by synchronizing their spikes in 40 Hz oscillations. These oscillations do not themselves encode additional information, except in so far as they join together some of the existing information into a coherent percept. We shall call this form of awareness "working awareness". We further postulate that *objects for which the binding problem has been solved are placed into working memory*. In other words, some or all of the properties associated with the attended location would automatically be remembered for a short time.

There are at least three experimental approaches to the study of working awareness: i) Look at *rivalry* as between equivocal perceptions of figures (for example, the Necker cube and the Schröder stairs in Figure 25) or binocular rivalry. These effects can be studied using the macaque monkey. ii) Allow cats to view the same pattern when asleep and awake and record from visual neurons. iii) Study the effects of anesthetics on awareness and recall in human subjects and on neuronal responses in monkeys.

In conclusion, Crick and Koch admit that theirs is not so much a detailed model of consciousness as a research program directed toward understanding the mechanism through which many items of information that are distributed about the brain can be so rapidly unified in perception. They see no need for "fancy quantum effects" in this program, and they suggest that much of the mystery of consciousness may disappear "when we can both construct such machines and understand their detailed behavior."

DANIEL DENNETT

Because it claims to cut the Gordian knot of the consciousness problem, *Consciousness explained* by Daniel C. Dennett has been widely discussed since its appearance in 1991. The author devotes a substantial portion of his book to stating the ideas that he does *not* support, and these include the following:

The mystification of consciousness. To Dennett this tendency, in a misguided effort to preserve the uniqueness and dignity of mankind, ignores the success that science has demonstrated in explaining the mysteries of genetic transcription and astrophysics.

The siren song of dualism. There is only one sort of stuff—the matter that is dealt with by physicists and chemists—and the dualism of Popper and Eccles is "deservedly in disrepute today" because the mind stuff (of Worlds 2 and 3) eludes physical measurement.

The Cartesian Theater. "The idea of a special center in the brain is the most tenacious bad idea bedeviling our attempts to think about consciousness." Cognition and control, in Dennett's view, are distributed about the brain rather than localized in some specific region.

Qualia. Such secondary qualities as colors, aromas, tastes, and sounds do not exist; they only seem to.

Since he has elected to avoid dualism "at all costs," Dennett is forced to deny the existence of a locus for consciousness (Cartesian Theater). In a longish chapter entitled "Qualia disqualified," he agrees that "There seem to be qualia, because it really does seem as if science has shown us that the colors can't be out there, and hence must be in here," but he finds the reasoning confused. To clarify matters, he imagines a robot that can discriminate between colors by storing pictures in which the various colored regions are labeled with numbers. From this model he claims, "There is no *qualitative* difference between the [robot's] performance of such a task and our own." Thus Dennett deals with James's Scylla and Charybdis by asserting that the monster and the whirlpool are identical.

As an alternative to the Cartesian Theater, he proposes a Multiple Drafts theory that is related to the parallel structure of Selfridge's Pandemonium architecture from the field of artificial intelligence that goes back to 1959 (Selfridge, 1959). On the experimental side, he proposes a method, which he calls "heterophenomenology," that would restrict the experimenter to the analysis of verbal texts.

The blindsight observations (Stoerig and Cowey, 1989; Weiskrantz, 1986) pose a problem for Dennett, and he devotes several pages to them. Beyond suggesting that the subjects might be "malingering" or "just pretending not to be conscious" (an attitude that is difficult to credit) he proposes experimental means for helping their responses become conscious—to some degree—of their underlying neural activities. Such proposals are interesting as clinical methods for switching the mechanism of consciousness back on, but it is difficult to see how this helps avoid the theoretical implications of the experiments.

In reading Dennett's book, one should be aware that his essentially functionalist position has been strongly criticized by several of his philosophical colleagues. Bernard Baars and Katharine McGovern (1993), for example ask, "Should we read his work for guidance on the fundamental issues? Are Dennett's ideas scientifically essential or incidental? Are they mainly heuristic? Or do they pose conceptual snares which productive science had better avoid?" Bruce Mangan (1993) sees it as resurrecting behaviorist dogma, and Josefa Toribio (1993) argues that "a better understanding is needed if Dennett's theory is to resolve, rather than sidestep, those problems related to the subjective phenomenology of consciousness."

I must agree. When Dennett asserts that there is no qualitative difference between the way that you and I remember a color and the way it is done by some imaginary robot, the engineer sighs. This is a striking example of Eccles's philosopher who bases his theory on "crude and antiquated information."

JOHN SEARLE

In his recent book *The rediscovery of the mind* (1992), John Searle argues that the terms of debate on the mind-body problem are outmoded and confused. The solution, he asserts, is simple and has been available "to any educated person since serious work began on the brain over a century ago." It is this:

> Mental phenomena are caused by
> neurophysiological processes in the brain
> and are themselves features of the brain.

He calls this point of view "biological naturalism" and asserts that "mental processes are as much [a] part of our biological natural history as digestion, mitosis, meiosis, or enzyme secretion."

Searle begins with an analysis of our recent intellectual history in an attempt to understand how it is that materialism seems to be the only rational way to approach the study of mind. Modern materialism appears in a variety of guises ranging from the claim that mental states don't exist (eliminative materialism) to the view that a computer that successfully mimics human behavior must have "thoughts, feelings, and understanding" (computer functionalism). Searle finds such attitudes to be implausible at best as he describes in detail the often unstated assumptions on which they are based.

Yet consciousness is—for each of us—a fact of reality that can be readily described. There is nothing strange about consciousness except that we don't understand it. It is a natural biological phenomenon, and its mystery today "is in roughly the same shape that the mystery of life was before the development of molecular biology or the mystery of electromagnetism was before Clerk-Maxwell's equations."

A section on emergent properties is particularly relevant to the point of view that is advanced in this book. Consciousness is said to be a "causally emergent" feature of the system of neurons in the brain in the same way that the properties of liquid water emerge from those of the H_2O molecule. Thus:

> The existence of consciousness can be explained by the causal interactions between elements of the brain at the micro level, but consciousness cannot itself be deduced or calculated from the sheer physical structure of the neurons without some additional account of the causal relations between them.

ERICH HARTH

Although many psychologists and philosophers subscribe to the orthodox materialism that was caricatured by William James—in the example of Shakespeare cited at the beginning of this chapter—and analyzed in detail by Searle, physical scientists often view this position with suspicion. Being familiar with the quantum theory, which denies predictability at the atomic level, and the theory of relativity, which mixes the concepts of space and time, physicists need not be overly impressed with philosophical conclusions that are based on scientific perspectives of the nineteenth century. A particularly effective example of such informed skepticism is presented by Erich Harth, whose recent book *The creative loop* (1993) treats the mind-brain problem in detail. As a physicist he is well situated to do so because he has studied the mechanisms of neurodynamics for many years.

With Karl Popper and Francis Crick, Harth chooses not to define consciousness, but he cites the blindsight experiments of Weiskrantz (1986) as evidence for a "mysterious monitor" that may become disconnected; thus—details apart— consciousness is a real phenomenon that is characterized by the following attributes: *Selectivity*, which is demonstrated by the fact that only a few of the myriad neural events are illuminated by consciousness, and these honored few seem to be arranged in some sort of hierarchy. *Exclusivity*, as exemplified by the the Necker cube and Schröder stairs in Figure 25. Thus consciousness deals with only a single perception at any moment. *Chaining*, which is the basic property of Hebb's phase sequence and exemplified by the passage from James Joyce in the previous chapter. Finally, *unitarity*. As the light of consciousness skips over the selected neural events, we feel ourselves to be the same person at this moment, yesterday, and at the time of our first memory.

Harth argues that positive feedback on a global scale plays a central role in the mechanism of consciousness. To understand positive feedback one must recognize that it differs in a fundamental way from negative feedback, its better-known cousin. Negative feedback is realized in the governor of a steam engine or a thermostat, and it has been proposed by the American mathematician Norbert Wiener (1961) as the basis for the science of *cybernetics* or "communication and control in man and machine." Negative feedback is essential for the design of the repeater amplifiers that are used in long-distance telephony because it stabilizes a system so the relationship between output and input (or effect and cause) is fixed (Bode, 1945).

Positive feedback, on the other hand, functions at the edge of instability, where the causal relationship between input and output is about to be lost. We have all heard this effect when the sound system in an auditorium breaks into spontaneous oscillation, or begins to *sing*. In such an event some slight sound is picked up by the microphone, amplified, and returned to the hall at a higher level that is self-sustaining. Generally such songs are undesired.

(I first met positive feedback, while in high school, as a ham radio operator. Being short of cash, I did not build my receiver using the almost universal superheterodyne principle but chose instead a super-regenerative design that required fewer parts. This little set had two knobs: one for tuning and the other for adjusting the positive feedback just below the level of spontaneous oscillation, at which point the set became extremely sensitive and selective. Starting with a scarcely audible signal, I could tune with one hand and adjust the level of positive feedback with the other until it became strong and clear, but constant monitoring and adjustment were required to keep the set just below the level of oscillation.)

Positive feedback is present in Hebb's concept of the cell assembly; this is what causes the cells to ignite as in the bonfire metaphor introduced in the previous chapter. Harth sees it as a means of *selective* amplification, and this is an important new idea. With reference to visual perception he proposes a feedback loop that includes higher cortical areas, the lateral geniculate nuclei (LGN), and the primary visual area: the *creative loop* of his title. In the context of the hierarchical structure of the brain, this positive feedback loop can be represented as in the following diagram where the arrows are intended to indicate causal interactions.

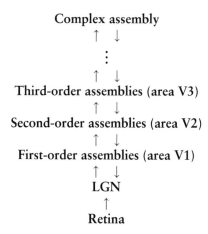

Complex assembly
↑ ↓
⋮
↑ ↓
Third-order assemblies (area V3)
↑ ↓
Second-order assemblies (area V2)
↑ ↓
First-order assemblies (area V1)
↑ ↓
LGN
↑
Retina

Such a loop remains in contact with the basic retinal data, which are topologically mapped onto the primary visual area (V1), so the connection between the original picture and its hierarchical representation within the neural jungle of the cortex is not lost. When the retina is not stimulated, activation of the same loop would lead to a mental impression of the corresponding visual pattern as in dreaming. Recent positron emission tomography experiments, showing neural activity in V1 when subjects were thinking about images, support Harth's theory (Kosslyn et al., 1993). Thus the entire loop sings, each level interacting with all the others in a dynamic entity of awesome complexity.

Viewed in this way, positive feedback is useful for the recognition of patterns that are imbedded in an irrelevant or deceptive context, like the A in the Danish letter Æ. Harth cites the example of finding a military tank that is hidden behind a tree in a scene containing clouds, trees, mountains, and a road: a task that is particularly difficult for a conventional computer code like the Perceptron mentioned in Chapter 6. Just as the careful adjustment of a super-regenerative receiver allows one to pick out, tune in, and amplify a weak signal among many others, the creative loop—guided by conscious attention—emphasizes those aspects of the scene that appear interesting (the tank), while progressively ignoring the others. This doesn't explain consciousness, as Harth admits, but it does suggest a mechanism that consciousness may employ in the course of perception.

The "process of exclusive selection" that is inherent in this mechanism, "is able to provide the unitarity of cognition that has been vainly sought by placing an intelligent monitor at the top of the sensory pyramid. But there is neither a theater there, as Dennett correctly points out, nor a homunculus observer to watch the plot. Dennett and others conclude from this that no unification takes place. Cognition is the production of multiple drafts; consciousness, a shared property of a horde of homunculi."

Harth concludes:

> These conclusions appear to be inevitable but challenge our intuition, which favors a unitary *I*. In the present, self-referent model that I have proposed, there *is* a theater, and the action on its stage is being scrutinized by an observer. Unlike previous attempts that have placed the theater at the highest level of cerebral activity, I believe that the unification is located at the only place where sensory patterns are still whole and preserve the spatial relations of the original scene—at the bottom of the sensory pyramid, not at the top. It is there that all the sensory cues and the cerebral fancies conspire to paint a scene. There is also an observer: it is the rest of the brain looking down, as it were, at what it has wrought. Consciousness, which arises in this self-referent process, not only unifies the immediate sensory messages but also becomes the joiner of everything around us, past, present, and future.

Dualism, according to Harth (1995), "is not quite as dead as some would have us believe," although this approach is often shunned as unscientific or antiscientific. In the context of human society, for example, a physical theory is nonsense; one must deal with social realities in order to understand why one man is made a king and another is hanged from a tree. Similarly, "such elements as intentionality and consciousness become valid concepts, no more mystifying, or science-defying than laws or politics were when we talked about society."

Finally, in comparing the brain with a digital computer, Harth notes that even the most powerful computer doesn't really think but performs a prescribed computational task in the service of a client. After these computations are complete, the client must examine them to determine what, if anything, they mean. The brain, on the other hand, is its own client.

HENRY STAPP

Henry Stapp, also a physicist, believes that quantum physics holds the key to an understanding of the mystery of consciousness. His arguments are based on a firm belief that classical physics implies reductive materialism. Thus he has little confidence in the dynamic complexities that can emerge from classical Newtonian dynamics. In his most recent work, *Mind, matter, and quantum mechanics* (1993), he asserts, "Nothing in classical physics can create something that is *essentially* more than an aggregation of its parts." This is in accord with the above-cited views of William James, whom he quotes extensively.

Since classical dynamics cannot be used to describe the properties of mind, it is necessary "in principle" to turn to Werner Heisenberg's formulation of quantum mechanics for an explanation of the properties of consciousness. The resulting description is called the Heisenberg/James theory.

There may well be "conglomerates that act cohesively as unified wholes," he admits, and such conglomerates may control their parts, but "to the extent that classical physics is valid" the dynamics of a material object can be controlled only by things that are themselves under deterministic control. The caveat—"to the extent that classical physics is valid"—is central to this argument. Classical mechanics is *not* valid, in Stapp's view, because it has been supplanted by quantum mechanics. Without quantum mechanics, "the evolution of the physical universe would be exactly the same whether subjective conscious experience exists or not."

The Heisenberg/James theory "identifies each conscious experience with a creative act." This act is represented in the physical world by the selection of one possibility from among the many that are generated by the dynamic evolution of quantum mechanics. How does this come about? From Heisenberg's perspective, there is a *wave function for the universe*, which evolves according to Schrödinger's equation and represents a set of "objective tendencies" or "propensities" for the occurrence of actual events. An actual event, in turn, is represented by a collapse of some facet of the universal wave function, and this collapse is governed by chance. Thus the universe is controlled in part by deterministic law (Schrödinger's equation) and in part by happenstance (the collapse events).

How is this new picture of reality incorporated into the dynamics of a brain? The place to begin, according to Stapp, is inside the synapse (see Figure 18), where calcium ions are released by an incoming action potential (nerve impulse) and diffuse to a transmitter site in a time that is short enough to require quantum corrections. If N synapses receive impulses within some small time interval and the probability of vesicle release is fifty percent, then there are 2^N possible configurations of vesicle release. "*Each alternative possibility is represented in the evolving quantum mechanical wave function.*"

Where, in the brain, does an actual event occur? Where does Heisenberg's universal wave function collapse in order to start off in a new direction of deterministic evolution?

John von Neumann has suggested that this collapse can be viewed as occurring anywhere on the hierarchy of dynamic activity. Assuming the collapse to happen at a particular level is called making the *Heisenberg cut* at that level. In the Schrödinger cat scenario, the reader will recall from Chapter 2, this cut is made at the level of a conscious observer as the box is opened.

From this perspective, the brain can be viewed as a hierarchical sequence of measuring devices for which the wave packet collapse (or Heisenberg cut between the quantum and classical pictures) can occur at any level: synapse, axon, neuron, or whatever. Stapp chooses the highest level of the brain's dynamic activity and associates making the cut with a conscious act. Thus "conscious events are assumed to be the feels of these top-level events, which actualize *macroscopic* patterns of neural activity."

I have several difficulties with this work. First of all, Stapp quotes liberally from the writings of Roger Sperry in support of his position, but this is curious because Sperry—as we have seen—believed the properties of mind to emerge from the complexities of the classical organization of the brain. Second, his assertion that "Nothing in classical physics can create something that is *essentially* more than an aggregation of its parts" is only that: an assertion. Like Roger Penrose, he fails to see the wealth of mystery that lurks in the bowels of classical theory. Perhaps he would argue that Jupiter's Great Red Spot or the Bosnian War is nothing more than the sum of its parts, but many would disagree. And neither is described by quantum theory. Third, the concept of a *wave function for the universe* is difficult for many—including our florist/chemist from Chapter 2—to take seriously.

In Chapter 9, I shall discuss some of the cultural and historical reasons that lead scientists to take such extreme positions.

ECCLES THROWS DOWN THE GAUNTLET

Strongly influenced by Stapp, John Eccles has recently taken up quantum theory (1992), and published a spirited defense of dualism in a book entitled *How the self controls its brain* (1994). Presented as a "challenge to all materialists," Eccles expresses sharp criticism of—among others—Crick and Koch ("science fiction of a blatant kind"), Dennett (an "impoverished and empty theory"), Searle ("a request for magic!"), and Sperry ("crude analogical arguments").

Throughout this book, Eccles is concerned with the problem of explaining the "self" or "soul" without violating the conservation laws of physics, in particular conservation of energy. How, he asks, can the self, which lives in World 2, influence the dynamics of the brain, which lives in World 1, without exchanging energy across the interface between the brain and the self, thereby breaking the law of energy conservation?

The philosophical position of Eccles is close to that of William James, but he goes further. In addition to postulating a soul to explain the evident facts of psychic reality, he proposes a mechanism by which it can influence the brain without exchanging energy and thereby violating energy conservation. This mechanism is based upon quantum theory, taking advantage of the fact that Schrödinger's wave function is a probability amplitude, which does not carry energy.

Together with Friedrich Beck, a theoretical physicist, Eccles has proposed that the release of transmitter molecules by a vesicle into the synaptic cleft (see Figure 18) is a quantum process (Beck and Eccles, 1992). Thus the probability of signal transmission across the cleft can be influenced by World 2 units of mental activity (called *psychons*) on World 1 bundles of cortical dendrites (called *dendrons*).

In the closing section of his argument, Eccles (1994) asks, "Will the materialists dare to reject it?"

CRICK REPLIES

As if in response to Eccles's challenge, Francis Crick published *The astonishing hypothesis* (1994a). In this book—subtitled *The scientific search for the soul*—Crick states the Astonishing Hypothesis:

> 'You,' your joys and your sorrows, your memories and your ambitions, your sense of personal identity and free will, are in fact no more than the behavior of a vast assembly of nerve cells and their associated molecules.

In support of this claim, Crick emphasizes the importance of emergent phenomena, a term that can be used in two senses. The first implies behavior that cannot be understood in any way and thus has mystical overtones. The second sense, which he uses, implies that "while the whole may not be the simple sum of the separate parts, its behavior can, at least in principle, be *understood* from the nature and behavior of its parts *plus* the knowledge of how these parts interact." Crick cites the organic molecule benzene, which is composed of six carbon atoms and six hydrogen atoms, as a simple example, and he points out that "nobody derives some kind of mystical satisfaction by saying 'the benzene molecule is more than the sum of its parts,' whereas too many people are happy to make such statements about the brain and nod their heads wisely as they do so."

In the context of the present book, the emergence of an electrochemical pulse on a nerve axon from the laws of physics and chemistry is emergence in Crick's second sense because an understanding of the process was developed in Chapters 2 through 4.

Written for the general reader, the book contains four chapters on the psychology of vision and seven that survey current knowledge about the structure of the visual cortex, which all but the most specialized reader will find both interesting and useful. Evidence for the hierarchical nature of visual processing is carefully reviewed, and from this perspective Crick concludes, "The secret of the neocortex, if it has one, is probably *its ability to evolve additional layers to its hierarchies of processing*, especially at the upper layers of those hierarchies." These extra layers of processing are thought to distinguish the more intelligent mammals from their slower cousins.

In an interview about the book with the *Journal of Consciousness Studies*, Crick (1994b) was asked his opinion of recently developed quantum mechanical models of the mind, and he replied as follows:

> Having spent a great deal of time following attempts to model single neurons, I know only too well about the different receptors and about vesicles and all the rest of it. We all know that quantum mechanics is the basis of chemistry, so no one can say that quantum mechanics isn't important. But what people are trying to say is that there is something more than chemistry involved. And I can't see that they have any grounds for that yet. Because they have not shown even in outline, not even as a sketch, what mysteries of the brain's function would be explained if something of this type did take place. So I am very skeptical of it.

The several threads of the book are drawn together in a final chapter entitled "Dr. Crick's Sunday morning service," in which it is again asserted that the time is now ripe to take the scientific problem of consciousness seriously. In this effort, he argues, it is essential to construct a theory that is based on neurons because *"The language of the brain is based on neurons."* He notes that this (the Astonishing Hypothesis) *is* a hypothesis, but one so plausible that many educated people do not find it astonishing.

Although I agree with most of what Crick says, the general reader may be confused because his initial statement of the Astonishing Hypothesis seems not to square with his subsequent description of it. To write that one's joys, sorrows, memories, personal identity, and free will are *no more than* the behavior of a vast assembly of nerve cells seems at variance with the statement that the brain—like a molecule of benzene—is *more than* the sum of its parts. His aim, clearly, is to divorce himself from the camps of both mystical dualism and quantum consciousness, but (like the wine buff who talks dry and drinks sweet) Crick states his position as a functionalist and then defends it as an emergent dualist.

THE HARD PROBLEM

Consciousness is a word with several meanings and it is used to describe a variety of mental phenomena, including being awake rather than asleep, having control of one's behavior, having a concept of one's self and of the world, and so on. The difficult aspect of consciousness, according to David Chalmers, an Australian philosopher, has to do with the *experience* of consciousness: our *awareness* of the ladder in the orchard, with which we began the introduction to this book. Thus Chalmers (1995) suggests that we divide questions of consciousness into those that are "easy" and "the hard problem": that of understanding the nature of conscious experience.

Take the practice of anesthesiology as an example. In preparation for surgery, your body is suffused with certain neurologically active molecules that bind to active sites of proteins in various regions of your brain, causing you to lose consciousness for a period of time, while your heart and lungs continue to beat and breathe. It is an interesting and important problem to learn which protein molecules are affected in what part of the brain and to understand how these changes at the lowest level of the brain's architecture manages to induce the global behavior that is described as "loss of consciousness." Medical science is a long way from having the answers to this problem.

Nonetheless, that is what Chalmers calls an "easy" problem. The related "hard" problem is to understand the nature of your experience of the operation: the mix of panic and resignation in your breast as you are transported along the antiseptic corridors, the intrusive glare of the lights, the ominous leer of that peculiar crack in the ceiling, the feeling of warm confidence that you receive from someone who touches your arm and looks into your eyes with love.

Although the "easy" problems seem difficult—and indeed *are* difficult— they ask for no change in the ways that scientists think. The answers to such problems are functional mechanisms of the sort that scientists are in the habit of prying out of their theories and laboratory experiments. The problem of conscious experience, according to Chalmers, requires a new way of thinking because it goes beyond questions of function to the essential nature of being.

Cognitive scientists have developed a variety of strategies to deal with the question of conscious experience. Some nod briefly in the direction of conscious experience and then explain something else, others deny its existence. Some claim to explain conscious experience but use theoretical tools that have logical gaps and seem to invoke magic, still others attempt to explain the structure of experience or isolate its substrate. None of these approaches succeeds, according to Chalmers, because an "extra ingredient" is always needed.

Extra ingredients abound. Nonlinear and chaotic dynamics is proposed as the panacea by certain applied mathematicians, while others propose nonalgorithmic information processing or point to new discoveries in neurophysiology. Quantum mechanics has recently become popular as an extra ingredient, and its attractiveness "may stem from a Law of Minimization of Mystery" according to which "consciousness is mysterious and quantum mechanics is mysterious, so maybe the two mysteries have a common source" (Chalmers, 1995). A special feature of quantum theory, as we saw in Chapter 2, is the role that consciousness is said to play in inducing the collapse of a quantum wave function, but how does this phenomenon explain the subjective nature of consciousness?

"At the end of the day," Chalmers concludes, "for any physical process we specify there will be a question that the physical theory leaves unanswered: Why should this process give rise to experience? For any such process, it remains conceptually coherent that it could be instantiated in the absence of experience. It follows that no mere account of the physical process will tell us why experience arises." In short,

You can't explain conscious experience on the cheap.

This, of course, is just the point being made by William James and George Santayana in the opening selections of the present chapter. How far, one wonders, has consciousness research progressed over the past century?

Chalmers recognizes that his position may seem like the claims of eighteenth-century vitalists that life was beyond physical explanation, but he sees the nature of life as a different question. It was not clear then—as it is now—that mere physical mechanisms could perform the functions of life. With consciousness, one goes beyond the physical explanation of functions. "The problem of consciousness is puzzling in an entirely different way, precisely because it is not a problem in the explanation of structures and functions."

References

B J Baars and K McGovern. Does philosophy help or hinder scientific work on consciousness? *Consciousness Cognition*, 2:18–27, 1993.

F Beck and J C Eccles. Quantum aspects of brain activity and the role of consciousness. *Proc. Natl. Acad. Sci. (USA)*, 89:11357–11361, 1992.

H W Bode. *Network analysis and feedback amplifier design*. D. Van Nostrand Company, Princeton, New Jersey, 1945.

D J Chalmers. Facing up to the problem of consciousness. In *Toward a science of consciousness*, S R Hameroff, A W Kaszniak, and A C Scott, editors. MIT Press, Cambridge, Massachusetts, 1995.

F Crick. *The astonishing hypothesis: The scientific search for the soul*. Simon and Schuster, New York, 1994a.

F Crick. Interview on *The astonishing hypothesis*. *J. Consciousness Stud.*, 1:10–17, 1994b.

F Crick and C Koch. Towards a neurobiological theory of consciousness. *Semin. Neurosci.*, 2:263–275, 1990.

D C Dennett. *Consciousness explained*. Little, Brown, Boston, 1991.

J C Eccles. Evolution of consciousness. *Proc. Natl. Acad. Sci. (USA)*, 89:7320–7324, 1992.

J C Eccles. *How the self controls its brain*. Springer-Verlag, Berlin, 1994.

R K C Forman. 'Of capsules and carts': Mysticism, language and the *via negativa*. *J. Consciousness Stud.*, 1:38–49, 1994.

E Harth. *The creative loop: How the brain makes a mind*. Addison-Wesley, Reading, Massachusetts, 1993.

E Harth. Self-referent mechanisms as the neuronal basis of consciousness. In *Toward a science of consciousness*, S R Hameroff, A W Kaszniak, and A C Scott, editors. MIT Press, Cambridge, Massachusetts, 1995.

D O Hebb. *The organization of behavior*. Wiley, New York, 1949.

D O Hebb. *Essay on mind*. Lawrence Erlbaum Associates, Hillsdale, New Jersey, 1980.

T Honderich. The time of a conscious sensory experience and mind brain theories. *J. Theor. Biol.*, 110:115–129, 1984.

W James. *Principles of psychology*. Holt, New York, 1890; republished by Dover, New York, 1950.

S Kosslyn, N M Alpert, W L Thompson, V Maljkovic, S B Wise, C F Chabris, S E Hamilton, S L Rauch, and F S Buonanno. Visual mental imagery activates topographically organized visual cortex: PET investigations. *J. Cognit. Neurosci.*, 5:263–287, 1993.

B Libet. Electrical stimulation of cortex in human subjects, and conscious sensory aspects. In *Handbook of sensory physiology*, A Iggo, editor. Springer-Verlag, New York, 1973.

B Libet. Subjective antedating of a sensory experience and mind-brain theories: Reply to Honderich. *J. Theor. Biol.*, 114:563–570, 1985.

B Libet, E W Wright, Jr, B Feinstein, and D K Pearl. Subjective referral of the timing for a conscious sensory experience. *Brain*, 102:191–222, 1979.

B Mangan. Dennett, consciousness, and the sorrows of functionalism. *Consciousness Cognition*, 2:1–17, 1993.

V W Mark. Conflicting communicative behavior in a split brain patient: Support for dual consciousness. In *Toward a science of consciousness*, S R Hameroff, A W Kaszniak, and A C Scott, editors. MIT Press, Cambridge, Massachusetts, 1995.

R Penrose. *The emperor's new mind*. Oxford University Press, Oxford, 1989.

R Penrose. Interview on *Shadows of the mind*. *J. Consciousness Stud.*, 1:17–24, 1994a.

R Penrose. *Shadows of the mind: A search for the missing science of consciousness*. Oxford University Press, Oxford, 1994b.

K R Popper and J C Eccles. *The self and its brain*. Springer-Verlag, Berlin, 1977.

G Santayana. *The sense of beauty*. Dover Publications, New York, 1955 (first published in 1896).

E Schrödinger. *What is life?*. Cambridge University Press, Cambridge, 1944 (reprinted 1967).

E Schrödinger. *Mind and matter*. Cambridge University Press, Cambridge, 1958 (reprinted 1967).

O Selfridge. Pandemonium: A paradigm for learning. *Brain*, in the proceedings of the *Symposium on the mechanism of thought processes*. H.M. Stationery Office, London, 1959.

J R Searle. *The rediscovery of the mind*. MIT Press, Cambridge, Massachusetts, 1992.

B F Skinner. *Beyond freedom and dignity*. Bantam/Vintage, New York, 1972.

R W Sperry. Forebrain commissurotomy and conscious awareness. *J. Med. Philos.*, 2:101–126, 1977.

H P Stapp. *Mind, matter, and quantum mechanics*. Springer-Verlag, Berlin, 1993.

P Stoerig and A Cowey. Wavelength sensitivity in blindsight. *Nature*, 342:916–918, 1989.

J Toribio. Why there still has to be a theory of consciousness. *Consciousness Cognition*, 2:28–47, 1993.

A M Turing. Can a machine think? *Mind*, 59:433–460, 1950.

J B Watson. *Behaviorism*. Norton, New York, 1924.

L Weiskrantz. *Blindsight: A case study and implications*. Oxford University Press, Oxford, 1986.

N Wiener. *Cybernetics*. The Technology Press and Wiley, Cambridge, Massachusetts, 1961.

E P Wigner. Are we machines? *Proc. Amer. Philos. Soc.*, 113:95–101, 1969.

"A Great Arc" of
Possibilities

*It is in our collective behavior that we are most mysterious.
We won't be able to construct machines like ourselves
until we've understood this, and we're not even close.*

Lewis Thomas

I n June of 1883, a fledgling German physicist sailed from Hamburg to spend a year on Baffin Island, between mainland Canada and Greenland. The purpose of his trip was to live with the native people and study their relationship to the surrounding geography. During the year, he came face to face with the subject of this chapter: the nature of a cultural configuration, and upon his return, he wrote:

> After a long and intimate intercourse with the Eskimo, it was with feelings of sorrow and regret that I parted from my Arctic friends. I had seen that they enjoyed life, and a hard life, as we do; that nature is beautiful to them; that feelings of friendship also root in the Eskimo heart; that, although the character of their life is so rude as compared to civilized life, the Eskimo is a man as we are; that his feelings, his virtues and his shortcomings are based in human nature, like ours. (quoted in Herskovits, 1953)

The young voyager's name was Franz Boas, and this Arctic experience changed the course of his life. He resolved to devote his professional career to the scientific study of mankind.

Boas's decision was a watershed event in the history of American social science. As a lecturer in the Department of Psychology at Clark University from 1889 to 1892, he supervised the first doctorate to be awarded in anthropology in the United States. Appointed to a lectureship in physical anthropology at Columbia University in 1896 and to a professorship in 1899, he created the first American Department of Anthropology, which he dominated until his retirement in 1937 at the age of 79 (Herskovits, 1953).

145

THE BIRTH OF AMERICAN ANTHROPOLOGY

In the late nineteenth century, American anthropology was a hobby practiced by amateurs, and—as both Melville Herskovits (1953) and Marshall Hyatt (1990) have emphasized—it was dominated by an evolutionary doctrine inherited from biology and steeped in racism. The degree to which a particular society had developed, most believed, could be measured by finding its place along a linear scale with the educated world of Europe and North America at the top and the most primitive tribe at the bottom.

Into this stale atmosphere, young Boas appeared like a spring breeze. Even before his trip to Baffin Island, he had begun to question the belief in reductive materialism that had dominated his studies in physics. Instead he turned to a general investigation of the interaction between inorganic and organic matter, and in particular the "relation between the life of a people and their physical environment" (Hyatt, 1990). "How far," he wondered in a question that is echoed in this book, "may we consider the phenomenon of life from a mechanistic point of view?"

In his lifelong search for an answer to this fundamental question, Boas understood two important features of complex dynamics a half century before they came to be appreciated by significant numbers of physical scientists. The first is *nonlinearity*, which he quaintly termed "differentiation" that leads to further differentiation. This effect, he wrote,

> ...may be observed in all specific phenomena of nature. A valley has been formed as the effect of erosion. It is the cause that in the further action of erosion the waters follow its course. Luxurious vegetation is the effect of a moist soil. It is the cause of retaining more moisture in the soil. A household performs joint work, and the joint work strengthens the unity of the household. Leisure obtained by preservation of a plentiful supply of food stimulates invention, and the inventions give more leisure. (Boas, 1928)

In each of these examples, one observes, the whole is more than the sum of its parts. The valley, the forest, the family, the community—each grows to become a robust entity like the tornado or the flock of geese that were described in Chapter 1 or the children on the trampoline from Appendix A.

Secondly, the crucial role that *historical accidents* play in the development of a culture was consistently emphasized by Boas throughout his long career, a perspective that is related to modern notions of *dynamic chaos*. Although genetic, economic, or geographic causes are sometimes proposed as determinants of the practices of mankind, Boas found such reductionist theories inadequate. One must not ignore the interrelated aspects of cause and effect in social dynamics nor neglect the importance of happenstance. Thus his classic *Anthropology and modern life* (1928) concludes with the statement:

> Accident cannot be eliminated, accident that may depend upon the presence or absence of eminent individuals, upon the favors

bestowed by nature, upon chance discoveries or contacts, and therefore prediction is precarious, if not impossible.

Thinking about our own lives—filled as they are with watershed decisions, chance meetings and unexpected tragedies—we must admit the truth of this perspective, though some choose not to do so.

One should not suppose that Boas's early rejection of reductive materialism led him toward beliefs in mysticism or divine intervention. Quite the contrary. Dedicated to the task of understanding "what determines the behavior of human beings," Boas brought from his training in physics a lifelong commitment to scientific accuracy and the "gift of skepticism" for theoretical ideas not based on carefully assembled data. According to Herskovits (1953), he gave to his science

> in addition to massive accumulations of facts, basic methodological and theoretical insights, while many of his critical analyses freed those who succeeded him from the need to explore paths which lead to propositions that would confuse, rather than clarify, the aims of the discipline. To the thinking of his time he gave a firm scientific support for tolerance toward racial and cultural differences, in terms so well reasoned and documented that much of what he stood for has moved into common thought, its source unsuspected by most of those who follow it.

Based on facts that he carefully observed and methodically recorded, Boas came to believe that anthropology should not aim to discover laws for the development of cultural configurations because such laws do not exist (Hyatt, 1990). Happenstance plays too strong a part. Instead the ethnologist should concentrate upon and attempt to understand the nature of the relationships between known cultural configurations and the individuals from which they emerge: the chemistry of human life.

CULTURAL CONFIGURATIONS

In our discussion of Donald Hebb's cell assembly in Chapter 6, the idea of a social assembly was introduced as a metaphor, but we are now ready to take the figure seriously. Our atoms are the conscious human minds considered in the previous two chapters, and the stage upon which they act is Karl Popper's World 3. The development of a particular cultural configuration from the interactions of individual personalities was described by Boas (1928) in the following terms:

> An isolated community that remains subject to the same environmental conditions, and without selective mating, becomes, after a number of generations, stable in bodily form. As long as there are no stimuli that modify the social structure and mental life the culture will also be fairly permanent.

As a molecule of water or benzene emerges from the nonlinear dynamics of the Born-Oppenheimer theory, as a nerve impulse emerges from the Hodgkin-Huxley equations, as a unique drama emerged in the mind of William Shakespeare, so does the culture of a community emerge from the collective experiences and interactions of its members.

Under the influence of Boas and his many brilliant students—including Ruth Benedict and Margaret Mead—cultural anthropology became a bona fide science of humankind rather than a collection of odd facts about primitive tribes. The central message of this science is presented in Benedict's classic *Patterns of culture* (1934), which compares the configurations of three quite different societies. From these comparisons we see that the story is much more complex than the relationship between molecules and atomic physics that we considered in Chapter 2. In her words:

> We must imagine a great arc on which are arranged the possible interests provided either by the human age-cycle or by the environment or by man's various activities. A culture that capitalized even a considerable portion of these would be as unintelligible as a language that used all the clicks, all the glottal stops, all the labials, dentals, sibilants, and gutturals from voiceless to voiced and from oral to nasal. Its identity as a culture depends upon the selection of some segments of this arc.

In terms of the present discussion, the number of possible languages is immense, and the number of possible cultural configurations that can organize human life is hyperimmense. As we saw in Chapter 6, humankind differs, in this respect, from other mammals. Why? Because the large ratio of association to sensory areas in the cortex of the brain, and the correspondingly long time to reach adulthood, make possible the emergence of a cultural configuration.

Just as a protein or an organism or a cell assembly or a tool is selected from "a great arc" of possibilities to fit its assigned task, a particular human culture takes a self-consistent form, a form that asks to be understood in its own context as an organic unity. Since the cultures of large modern nations are both dynamic and extremely complex, it helps to begin with studies of simpler and more stable societies from which one may obtain some degree of fundamental understanding.

The three cultures described by Ruth Benedict are these:

The Zuñi of New Mexico. No one who visits the graceful cliff dwellings at Mesa Verde or Bandelier National Monument or the grand circular cities in Chaco Canyon can fail to be impressed by the sense of order and balance in the culture of those who built them. Their descendents—the Zuñi—are described as an Apollonian people "whose delight is in formality and whose way of life is the way of measure and sobriety." Their ceremonial life is (or was when Benedict wrote) a daily preoccupation directed primarily toward the inducement of rain and growth of the tribe and its gardens. Such matters as marriage and divorce—which consume so much time and energy in our culture—are arranged without fanfare and with a minimum of formality. Homicide is almost unknown and anger rare among the Zuñi, and hallucinations are feared. They are repelled by drunkenness and unable to conceive of suicide.

Benedict quotes D. H. Lawrence, the novelist and sometime resident of the American Southwest, who observed that in a Zuñi dance "All the men sing in unison, as they move with the soft, yet heavy bird tread which is the whole of the dance, with bodies bent a little forward, shoulders and heads loose and heavy, feet powerful but soft, the men tread the rhythm into the centre of the earth. The drums keep up the pulsating heart beat and for hours, hours, it goes on."

Leadership is not sought but grudgingly accepted, and strong runners are excluded from foot-races in order not to spoil the game. The Zuñi have no sense of sin, for the idea of evil is foreign to them. "The breath of the gods is their breath, and by their common sharing all things are accomplished."

The Dobu of southeastern New Guinea live on barren islands with poor soil and little fishing. Formerly cannibals, they were rightly feared by their neighbors as lawless, treacherous, and diabolic. Within the culture each person views his or her daily associates as sorcerers who pose a constant threat. This hostility extends into marriage, which begins with the entrapment of the groom by his future mother-in-law in her daughter's bed and leads to a year of humiliation and enforced labor. Relations of each spouse with his or her in-laws are so unpleasant that the pair must alternate yearly between two villages. Although adultery is common—even expected—its discovery causes violent outrage leading to divorce or a sullen and angry marriage.

In the Dobu culture, husband and wife fear each other and care for their yam gardens separately because neither will trust the other with the magic that makes them grow, and failed crops lead to divorce. The Dobu are intensely competitive over material possessions and sex and especially for control of the magic that is thought to govern all that transpires. Life, for them, is a zero-sum game; for every winner there must be a loser. To injure someone, a man feigns friendship in order to find the opportunity for treachery, and murder—even of children—is common. "If we wish to kill a man we approach him, we eat, drink, sleep, work and rest with him, it may be for several moons. We bide our time. We call him friend."

The Kwakiutl of Vancouver Island. In contrast to the Zuñi, the Indians who lived on the northwest coast of America, between Alaska and Puget Sound, were Dionysian, and, unlike the Dobu, they were wealthy. Their coast was dotted with islands and the sea teemed with fish. Wood was easily taken from the nearby forests so—in addition to hunting and fishing— the men occupied themselves with woodworking, building fine houses and handsome seagoing canoes. Their art was striking and—like the majority of Native Americans—they strove for ecstasy in their religious practice.

Benedict's account, based on the field work of Franz Boas published between 1898 and 1930, focuses on the Kwakiutl of Vancouver Island, for these had a unique social configuration. Although—or perhaps because—the eating of human flesh was repugnant to the Kwakiutl, the Cannibal Dance was an important part of the Winter Ceremonial. Caught in the act of eating human flesh, the Dancer was finally brought under control by the enticements of a naked woman carrying a corpse. Apart from their taste for extreme experience, the Kwakiutl viewed life as struggle between triumph and shame where "Triumph was an uninhibited indulgence in delusions of grandeur, and shame was a cause of death." This attitude colored every aspect of their lives. Dinners between chiefs became *potlatches* in which the host and his most honored guest tried to shame each other by giving away or destroying more property than the other could afford. Marriage—so simple among the Zuñi—was a vast and complex economic competition, spanning decades. When his self-esteem was damaged, the Kwakiutl would sulk and consider suicide, and to deal with a loved one's death he would transfer the pain to another. A chief whose son had died visited a neighboring chief, and after being cordially received, said, "My Prince has died today, and you go with him," and then killed him. He assumed the same attitude toward his gods, cursing them when they failed.

Although the examples chosen emphasize the importance of an integrated cultural configuration, Benedict recognized the variability of this trait, noting that cultures may be more or less integrated just as individual personalities may. The notion of society as an organism, she emphasized, is not a surrender to animism or mysticism—as some would have it—but a recognition that individual psychology is unable to explain the facts of cultural integration.

COMING OF AGE IN SAMOA

In the spring of 1923, a young student at Columbia University was discussing with Ruth Benedict whether she should pursue her doctoral studies in psychology or in sociology. When Benedict said to her, "Professor Boas and I have nothing to offer you but an opportunity to do work that matters," the question was settled. Margaret Mead chose cultural anthropology (Mead, 1972). Two years later she was on her way to the island of Samoa, where—alone and 2 years younger than her mentor was on Baffin Island almost 40 years before—she would spend 9 months learning to be a field ethnologist. Since most of social anthropology to that time had involved men studying men, the task that Boas set for her was to investigate female adolescence, and she responded with bravura. Her book on the experience—*Coming of age in Samoa* (1928)—written in a direct style without professional jargon for the general reader, became a scientific classic. (Although certain revisionists questioned the reliability of some of her data in the early 1980s, it is difficult to understand why one should not expect some cultural alteration after a half century of exposure to European missionary schooling and a world war [Howard, 1984].)

Mead observed that female adolescence in Samoa was not the period of intense mental and emotional stress that one then expected in the United States, and indeed still does. "What is there in Samoa," she asked "which is absent in America, what is there in America which is absent in Samoa, which will account for this difference?" First of all, she noted the general casualness of Samoan society, where strong convictions were few and disagreements easily settled. Failure and defeat were not considered to be great disasters, and the Samoan society rewarded those who took them lightly.

Equally important, in Mead's view, was the number of choices open to young people in our society compared with those available to Samoans. Just as the molecule of the protein myoglobin goes "kicking and screaming" through its chaotic dynamics rather than exhibiting the peaceful quasi-periodic motions of a water molecule, just as the brain of Erwin Schrödinger was more tempestuous than that of his cat, a large, modern culture is far more difficult to comprehend than an isolated collection of rural villages. In a memorable passage, Mead described the complexity of American culture in the following words:

> Our young people are faced by a series of different groups which believe different things and advocate different practices, and to each of which some trusted friend or relative may belong. So a girl's father may be a Presbyterian, an imperialist, a vegetarian, a teetotaler, with a strong literary preference for Edmund Burke, a believer in the open shop and a high tariff, who believes that a woman's place is in the home, that young girls should wear corsets, not roll their stockings, not smoke, nor go riding with young men in the evening. But her mother's father may be a Low Episcopalian, a believer in high living, a strong advocate of States' Rights and the Monroe Doctrine, who reads Rabelais, likes to go to musical shows and horse races. Her aunt is an agnostic, an ardent advocate of woman's rights, an internationalist who rests all her hopes on Esperanto, is devoted to Bernard Shaw, and spends her spare time in campaigns of anti-vivisection. Her elder brother, whom she admires exceedingly, has just spent two years at Oxford. He is an Anglo-Catholic, an enthusiast concerning all things mediæval, writes mystical poetry, reads Chesterton, and means to devote his life to seeking for the lost secret of mediæval stained glass. Her mother's younger brother is an engineer, a strict materialist, who never recovered from reading Haeckel in his youth; he scorns art, believes that science will save the world, scoffs at everything that was said and thought before the nineteenth century, and ruins his health by experiments in the scientific elimination of sleep. Her mother is of a quietistic frame of mind, very much interested in Indian philosophy, a pacifist,

a strict nonparticipator in life, who in spite of her daughter's devotion to her will not make any move to enlist her enthusiasms. And this may be within the girl's own household. Add to it the groups represented, defended, advocated by her friends, her teachers, and the books which she reads by accident, and the list of possible enthusiasms, of suggested allegiances, incompatible with one another, becomes appalling.

How does one make a science of predicting the paths that the young will choose in such a complex culture?

CULTURAL RELATIVITY

If Wilder Penfield's patient (M.M.) had been born into an Indian tribe of California, for example, or medieval Europe, her tendency to have experiential hallucinations might have led her to become a mystic rather than to endure the surgical procedure shown in Figure 26. The clergy of Puritan New England would be treated as psychoneurotics today, and the women they tormented and murdered might be considered normal. This is not to accept a mindless relativity in which all cultural configurations are equally valid; there are some measures of cultural value, but even for Boas this was a tricky question. As a scientist, according to Herskovits (1953), he found science good. Conditioned by his culture, like those he studied, he was moved by the vision of a science that would help mankind overcome the ageless problems of poverty, war, and servitude. Unwilling to see science as but one of many responses to the challenges of nature, he was prepared to measure social progress as the "development of invention and knowledge," although "other aspects of cultural life are not with equal ease brought into a progressive sequence."

How *does* a tree manage to visualize its forest? We have met this problem before in connection with Schrödinger's attempt to describe consciousness. Since the sentient self is identical with one's world picture, it is difficult to view it objectively. Popper suggested that an understanding of the relationship between Worlds 3 and 2 might help one to see a corresponding relationship between Worlds 2 and 1. Perhaps a cultural configuration emerges as the *collective consciousness* of a people.

Although there is ever the danger of myopia in considering one's own culture, we should at least try to judge our dominant traits. Ruth Benedict believed a culture that provides opportunities for unfettered development of the most diverse of its individuals should be considered better than one that forces most of its members to conform to a narrowly defined personality or suffer the pain of being an outsider. This requires both a general recognition of cultural relativity and tolerance for individual diversity:

> We shall arrive then at a more realistic social faith, accepting as grounds of hope and as new bases for tolerance the coexisting and equally valid patterns of life which mankind has created for itself from the raw materials of existence. (Benedict, 1934)

ANTHROPOLOGY AND MODERN LIFE

Based on a half-century of dedicated and original research in both physical and cultural anthropology, *Anthropology and modern life* (1928) by Franz Boas established modern currents of thought on several important social problems. His approach is scientific rather than polemic—with a careful regard for reliable data—so the book continues to be studied by succeeding generations of social scientists.

Anthropology as defined by Boas is the science of mankind with emphasis on the group rather than the individual, and from this perspective the concept of race is introduced in a manner that emphasizes the insignificant effects of racial differences on cultural life. Race consciousness does exist, of course, but it springs from an almost universal and perhaps genetic tendency to form closed societies—or *social sets*—in complex cultures, and it varies greatly from one culture to another.

Nationalism (or tribalism or factionalism), eugenics, and criminology are considered in turn, and toward each an appropriate scientific attitude is developed. From a wide variety of examples, Boas demonstrates that the sense of nationality is but weakly related to a common race, language, or culture. Since eugenic selection can enhance only those features that are heritable and the statistical arguments of cacogenics are unconvincing at best, it is prudent to assume that complex activities are socially determined unless the contrary can be unequivocally demonstrated. What constitutes a criminal act, for example, clearly depends on what is regarded as a crime: a truism that carries one's thoughts back to Alan Turing. (Shortly before he took his life, this brilliant scientist had been forced by benighted representatives of justice to undergo hormone therapy for a sexual orientation that was then deemed criminal [Hodges, 1983].)

Just as the statistical fact that relatives of farmers tend to be farmers is not to be taken as evidence that the agricultural life is hereditary, the irrelevance of environment to criminal activity has not been shown. Although this seems evident, one continues to read in reliable journals of schemes to solve the crime problem on the cheap by sorting or controlling individual genetic codes. And proposals again appear to promote reactionary social policies based on dubious interpretations of IQ measurements for different races. The attitude of Franz Boas toward such uses (misuses, really) of social data is entirely in accord with the findings of Donald Hebb and his colleagues in the postwar decades. As we have seen in Chapter 6, this work clearly demonstrated the importance of a rich variety of social experiences for intellectual growth. With characteristic clarity, Hebb commented somewhere that to argue about whether nature or nurture is more important for the development of adult intelligence is like asking which contributes more to the area of a rectangle: its length or its width.

Boas showed that our particular economic system cannot be explained as an innate human competitiveness nor war as a biological need to fight. Such traits are not related to heredity. The Apollonian Zuñi have the same genetic background as the Dionysian Indians of the plains, and the European gene pool has not changed greatly from the Middle Ages to the present (Benedict, 1934). In his words:

> Many examples can be given showing that people of essentially the same descent behave differently in different types of social setting. The mental reactions of the Indians of the western plateaus, a people of simple culture, differ from those of the ancient Mexicans, a people of the same race, but of more complex organization. The European peasants differ from the inhabitants of large cities; The American-born descendents of immigrants differ from their European ancestors; the Norse Viking from the Norwegian farmer in the northwestern States; the Roman republican from his degenerate descendents of the imperial period; the Russian peasant before the present revolution from the same peasant after the revolution. (Boas, 1928)

Human cultures developed slowly in the beginning, but recently there has been a general tendency for change to accelerate. Simple cultures of today are relatively stable if insulated from outside influences, but complex cultures can be disturbingly unstable. The source of this instability, as Margaret Mead noted, is not only the pressure of technological change but also the fact that the many subcultures of a complex culture provide such a wide variety of patterns for individuals to follow.

As we race headlong toward one of the many possible futures that awaits us, do we have any means of control? Do we, as a culture, know that we have a choice? How shall we guide our collective canoe through the white waters of the future? How do we begin to consider such questions?

ETHICS

At this point, we stray beyond the purview of science for a brief consideration of ethics. This subject was raised by Schrödinger (1958), who observed that for an organism to live in a community it must forgo the egotism of the solitary and practice some degree of altruism, and it is closely related to the theory of values that George Santayana described as the basis of aesthetics. The sense of goodness is akin to the sense of beauty; both stem from the conscious mind.

"Altruism" is also the title of the last chapter of Charles Sherrington's classic *Man on his nature* (1951). To appreciate Sherrington's approach to altruism, it is important to recognize that he starts from a profound belief in natural science. Although many religions proclaim a respect for living creatures and examples of symbiotic cooperation are often observed, much of nature's activity involves fierce struggles for existence that are notoriously wasteful of life. To the question "Is this bad?" she provides no answer. She doesn't entertain such questions, but coldly considers evolutionary competition as "a favorable field for development in the scramble for a means of living." It is when mind enters the picture that suffering begins.

In the conflicts of life, consciousness helps both those who attack and those who choose to defend themselves. An attractive strategy is the defensive community; thus in Sherrington's words:

> The herd, the flock, the hive, groups of related individuals with interests in common, organized on a peace footing, the individual as unit contributing social safety and support to the community of units. Under this organization the mind gradually evinces new qualities of the 'self.' Zest-to-live takes on new aspects. Thus, in our own humankind altruism, extending to the family and beyond, knitting social ties of planet-wide comradeship and good will. Love-of-life extends, so to say, to beyond 'self.' It is sublimated to new aspirations, which in their fullness grow strong and dear as love-of-life itself—pity and charity and love of others and self-sacrifice, even to the sacrifice of 'love-of-life' itself.

On the other side of the coin is the predatory mode of existence, which Sherrington suggests also has its attractions.

> It has prospered and has prospered hugely. It has brought into existence and maintains in existence countless millions of lives which otherwise could not have been and but for it would forthwith perish. It has produced beautiful types of form and motion. It attaches as corollary to 'zest-to-live' a 'lust-to-kill.' It develops its own gift of killing to heights of skill and ingenuity which astonish. The predatory type of life even where endowed with developed mind is more often a noncommunity life. It brings, and this may be regarded as inherent in it, little progress in social organization.

Nature—being amoral—does not consider life sacred, but suffering increases with the emergence of mind as one ascends the evolutionary scale. Mankind is both cursed and blessed with the zest-to-live. In the light of natural science it is no longer possible to look outside for guidance in these matters. Thus:

> We have, because human, an inalienable prerogative of responsibility which we cannot devolve, no, not as once was thought, even upon the stars. We can share it only with each other.

BIOLOGICAL AND CULTURAL EVOLUTION

During his long and influential career in American science, Franz Boas was not without critics and detractors. Foremost among these was Charles Doolittle Walcott, the preeminent paleontologist of his day, who discovered the fossil data from the Burgess Shale in the Canadian Rockies. According to Stephen Jay Gould (1989, 1994), dozens of different phyla suddenly appeared during the Cambrian "explosion" (about 530 million years ago), most of which were lost in the post-Cambrian "decimation"—a dramatic example of the influence of happenstance in the course of biological evolution. Yet Walcott attempted to "shoehorn" these data into the preferred structure of an evolutionary tree. Why? Because the tree is an expression of evolutionary inevitability, and the reductive mind has a subconscious need to predict.

In one of the less savory chapters of scientific history, Walcott strove to drive Boas out of American science during the First World War (Gould, 1989; Hyatt, 1990). Boas's crusty inflexibility in matters of scientific integrity and Walcott's chauvinism and anti-Semitism suffice to explain the rancor of the dispute, but I suspect that the difference in their scientific perspectives—Walcott's reductive materialism vs. Boas's concern for happenstance in the historical record—was also a factor.

More recently some of Boas's arguments have been challenged by proponents of *sociobiology*, a branch of science that tends to overemphasize the importance of genetic endowment in the development of human culture (Wilson, 1980). This lapse into a reductive view of ethnology has been rebutted by Tim Ingold, an English social anthropologist, who argues that organisms are not machines—constructed piece by piece—because they *grow*. Similarly, a culture is not merely a collection of unrelated genetic traits, but an integrated configuration, having developed so over the course of its history. In his words:

> We do not posit individuals in advance as ready-made, functioning entities, and generate social life by imagining them to associate and to interact under the impulsion of their separate natures. We rather *start* with social life, as a progressive 'building up' of relationships into the structures of consciousness. This 'building up'...is equivalent to the generation of persons. (Ingold, 1990)

Social life, from this perspective, "is a process, consisting in the creative unfolding of relationships and the becoming of persons." And

> a theory of persons can be encompassed within a more general theory of organisms, without compromising the role of human agency or denying the essential creativity of human life. This creativity, magnified a thousandfold by the work of consciousness, is but a specific aspect of the universal capacity of organisms to act, in a certain sense, as the originators of their own development.

Recalling again Karl Popper's suggestion that we can learn about the relationship between the brain and the mind by studying the relationship between a cultural configuration and the persons who comprise it, we see that consciousness emerges as the mind develops in two opposite directions. Although it comprises the unique experiences of an individual, consciousness is no less a component of the cultural configuration in which it develops. The individual's genetic code (carried by the molecules of DNA that reside in the nuclei of the cells and fix the primary sequences of his or her proteins) is clearly an important aspect of the process, but the recognition of this fact should not blind us to the equal importance of other factors. The process of growth must also be recognized. Although the DNA code circumscribes the "great arc of possibilities" for human development, that arc is hyper-immense. We shall never exhaust its possibilities for human organization.

Having climbed the stairway to the mind, step by step, from biochemistry to electrophysiology to psychology to ethnology, we see that social reality is far too rich and far too subtle to fit comfortably within the narrow confines of genetic reductionism.

References

R Benedict. *Patterns of culture*. First published in 1934, and republished by Houghton Mifflin, Boston, 1989.

F Boas. *Anthropology and modern life*. First published in 1928, and republished by Dover Publications, New York, 1986.

S J Gould. *Wonderful life: The Burgess Shale and the nature of history*. W.W. Norton & Company, New York, 1989.

S J Gould. The evolution of life on earth. *Sci. Am.*, 271 (No. 4):84–91, October 1994.

M J Herskovits. *Franz Boas: The science of man in the making*. Charles Scribner's Sons, New York, 1953.

A Hodges. *Alan Turing: The enigma*. Simon & Schuster, New York, 1983.

J Howard. *Margaret Mead: A life*. Simon & Schuster, New York, 1984.

M Hyatt. *Franz Boas: Social activist*. Greenwood Press, Westport, Connecticut, 1990.

T Ingold. An anthropologist looks at biology. *Man* (N.S.), 25:208–229, 1990.

M Mead. *Blackberry winter: My earlier years*. William Morrow & Company, New York, 1972.

M Mead. *Coming of age in Samoa*. First published in 1928, and republished by William Morrow and Company, New York, 1973.

E Schrödinger. *Mind and matter*. Cambridge University Press, Cambridge, 1958 (reprinted 1967).

C S Sherrington. *Man on his nature*, Second edition. Cambridge University Press, Cambridge, 1951.

L Thomas. *The Lives of a cell*. Viking Press, New York, 1974.

E O Wilson. *Sociobiology: The new synthesis*. Harvard University Press, Cambridge, Massachusetts, 1980.

Toward an Emergent
Theory of Consciousness

I came upon a vulture high
Upon a lonely mountain peak.
He stared at me with silent eyes,
I looked at him afraid to speak...

Then faced him as the world grew still
Until did he his wings unfold
And disappear beyond the hills
To leave me standing in the cold.

Alwyn Scott (previously unpublished)

In the preceding chapters, I have tried to lay down two broad paths. The first winds through the scientific evidence relating to the investigation of consciousness: this path, unfortunately, becomes lost in an impenetrable thicket deep in the brain, where most neurobiologists will admit they do not yet know how or where consciousness arises.

The second path concerns the philosophical and theoretical approaches to the problem of consciousness: the exploration of what consciousness may prove to be. This road is paved with ancient stones as well as modern ones. But no single volume, to be sure, can possibly mention every one.

In attempting to describe the major landmarks along each path—especially those erected in the past century—I have introduced the reader to the length and variety of the discussion. Do these two serpentine paths intersect? Yes, they do, but before showing how, I must make my own position clear.

My personal views on the intersection between the science and philosophy of consciousness have been formed by interactions with many able researchers and theorists. Thus I cannot claim any great originality for my own opinions. Nonetheless I can state without qualification that I do not believe consciousness can be analyzed in the same way that a hydrogen atom can be understood. Indeed, as I hope the earlier chapters of this book have demonstrated, consciousness is an awesomely complex phenomenon. It is so complex that it cannot be reduced to some fundamental theory or to one simple biological or chemical transaction. Instead it must be approached from every level of the hierarchy. We must construct consciousness from the relevant physics *and* biochemistry *and* electrophysiology *and* neuronal assemblies *and* cultural configurations *and* mental states that science cannot yet explain.

In taking an emergent view of consciousness—a view in which the phenomenon must be understood by examining its many layers—I fully recognize that it is not an entirely orthodox or widely held position. Reductive materialists, in fact, insist on precisely the opposite view. And for many well-entrenched adherents of the reductionist school, with a fair degree of success in other scientific quests, it is difficult to concede that their swords are perhaps too blunt to cut consciousness down to size. Those who support a hierarchical perspective are fewer, if no less persuasive, but are less able to draw upon three centuries in the annals of scientific history to buttress their approach.

Part of the hesitancy to abandon the reductive tradition, I suspect, is rooted in a fear of disrupting certain articles of faith that go back many centuries. These articles, to me, seem rather similar to religious credos that most scientists have long since abandoned in their professional lives. Suppose, for instance, that I were to announce at a large scientific meeting that I believed in an omniscient God, to Whom every detail of the past, present, and future is known. I suspect that my scientific colleagues would write me off as a misguided fundamentalist.

But now consider another possibility. If I were to profess belief in a theory of everything, a vast body of knowledge that determines every fact of the future from the facts of the past, many would welcome me as a comrade—a man loyal to the tenets of science. I would not be given another glance; the ranks of science include many such people. There may be a valid distinction between the two positions, but it eludes me.

The roots of the bias toward determinism may lie deeper in our cultural history than scientists are accustomed to suppose. Indeed, it is possible that they may *predate* modern scientific methods. In his analysis of thirteenth-century European philosophy, Henry Adams (1904) observed:

> Under any conceivable system the process of getting God and Man under the same roof—of bringing two independent energies under the same control—required a painful effort, as science has much cause to know.

Archly he added, "Saint Thomas did not allow the Deity the right to contradict himself, which is one of Man's chief pleasures." One wonders to what extent reductive science has merely replaced Thomas's God with the theory of everything.

More recently, however, certain physicists have begun to challenge the reductionist approach to biological knowledge (Schweber, 1993). In this debate, the emergent view was well expressed twenty-five years ago by Walter Elsasser, a theoretical physicist who developed ideas about biology that had previously been suggested by Niels Bohr (Bohr, 1933; Elsasser, 1969a, b). Since the immense numbers of possible structures at each level of the biological hierarchy far exceed the number of organisms that actually exist, it is difficult to develop biological laws by averaging over identical individuals. Organisms were said to be "radically inhomogeneous" because "they contain structure within structure within structure, at any level from the grossly macroscopic to the molecular one. This suggests that different configurations in very small dimensions, that is at the molecular level or slightly above, might after some time eventuate in clearly distinguishable macroscopic effects." In a very real sense, this book is an attempt to develop further Elsasser's ideas.

A similar view was expressed at about the same time by Philip Anderson (1972), a condensed matter physicist, who argued that

> the reductionist hypothesis does not by any means imply a 'constructionist' one: The ability to reduce everything to simple fundamental laws does not imply the ability to start from those laws and reconstruct the universe. In fact the more the elementary-particle physicists tell us about the nature of the fundamental laws, the less relevance they seem to have to the very real problems of the rest of science, much less to those of society. The constructionist hypothesis breaks down when confronted with the twin difficulties of scale and complexity.

Anderson pointed out that the sciences can be arrayed in the hierarchy shown in Table 2 where the elementary entities of science X obey the laws of science Y. At each level of this hierarchy, "entirely new laws, concepts and generalizations are necessary." Indeed, I may have gotten the idea of the scientific hierarchy from reading Anderson's paper; it is too long ago to recall.

The physicist Erich Harth (1995), one of the ablest defenders of the emergent credo, has written that the position of the reductionist camp is one that is logical and plausible mostly to its own members. According to Harth,

> To say that all of human affairs is describable and explainable in strictly physical terms is sheer nonsense. It is equally nonsensical to assert that introducing such elements as political philosophies,

X	Y
Many-body physics	Elementary particle physics
Chemistry	Many-body physics
Molecular biology	Chemistry
Cell biology	Molecular biology
•	•
•	•
•	•
Psychology	Physiology
Social sciences	Psychology

Table 2. Anderson's scheme for the hierarchical structure of science.

or laws, or a *climate of opinion*, means resorting to some kind of mysticism and embarking on a non-scientific cabal. We cannot expunge such concepts from a discussion of societal dynamics, unless we confine ourselves to describing patterns of movement of people through subway turnstyles during rush hour. It must be apparent to all but the most simple-minded reductionist that the attempt to construct a true physical theory of society would be a foolish undertaking.

Harth is not alone. In *The quark and the jaguar* (1994), the theoretical physicist Murray Gell-Mann finds the concept of a theory of everything to be

a misleading characterization unless 'everything' is taken to mean only the description of the elementary particles and their inter-actions. The theory cannot, by itself, tell us all that is knowable about the universe and the matter it contains. Other kinds of information are needed as well.

Those who disagree should either prove the claim of reductive determin-ism, admit that it is an assumption, or give it up. Surely this is not too much to ask of scientists. Doubting the claim does not constitute a belief in magic, because magic implies the contravention of known laws of exact science—as with precognition, spoon bending, and the like. Many things that I don't understand manage to obey the laws of physics and chemistry.

CREDO

I feel comfortable with the view that consciousness—like life—is a real phe-nomenon. There are at least three reasons for making this assumption: first of all the split-brain (Sperry, 1977) and blindsight (Weiskrantz, 1986) ex-periments, among others, provide objective evidence for the existence of a mental monitor that might or might not be in operation during a particular act of visual response. Secondly, consciousness gives an evolutionary advan-tage to the species that develops it. Finally, personal introspection provides evidence that cannot be denied. These reasons suffice to convince me that it is imprudent—indeed unscientific—to assume that consciousness does not exist merely because we lack an explanation for it.

My perspective on the nature of consciousness is closely related to the structure of the hierarchy of knowledge that was sketched in the Introduc-tion, and it is interesting to see how neatly the formalism of Popper and Eccles (and ultimately that of Plato) is in accord with this scheme: World 2 is the level of consciousness, World 1 is everything below, and World 3 is everything above.

Although this seems to be a dualist position, I do not share the concerns about the mind-brain problem that were expressed by Popper and Eccles. Just because we can trace the nonlinear diffusion equations of Hodgkin and Huxley back to the laws of chemistry and physics does not mean that the laws of chemistry and physics *imply* the Hodgkin-Huxley equations. These equations do not disagree with the laws of physics and chemistry, but several hierarchical levels and combinatorial barriers prevent such a derivation. Thus, as was noted at the close of Chapter 4, the Hodgkin-Huxley equations can be viewed as "new laws" that are independent of atomic physics.

Since the dynamical laws governing consciousness are about as far removed, in a hierarchical sense, from the all-or-nothing dynamics of a nerve impulse as the Hodgkin-Huxley equations are from those of physics and chemistry, a similarly obscure relationship is expected to obtain between the Hodgkin-Huxley equations and consciousness. Thus it is misleading to try to imagine what the brain might do in terms of the dynamics of electrochemical pulses. To put this as clearly as possible:

> The dynamic laws that govern consciousness are no more closely related to the all-or-nothing action of a nerve fiber than the Hodgkin-Huxley equations are to the laws of atomic physics.

Although I admit that I am not able to explain consciousness, this admission is not evidence of a benighted effort to preserve the "freedom and dignity" or the "uniqueness and dignity" of humankind—as disciples of Skinner or Dennett might claim. It is simply a recognition that the reality of mind is complicated far beyond the present reaches of our collective imagination.

QUANTUM CONSCIOUSNESS AND SCIENTIFIC DETERMINISM

As the comments on Schrödinger's cat in Appendix E would suggest, I agree with Francis Crick that there is no need to invoke "fancy quantum effects" to explain the phenomena of consciousness. What is the basis for this point of view?

Recent assertions that quantum theory must play a fundamental role in mental dynamics seem to stem from the lingering influence of Laplacian determinism in our Western science. The future is thought to be determined from the present by scientific law, and according to William James "a clearheaded atomistic evolutionist" must conclude that "nothing... but the everlasting atoms" plays a role in the dynamics of William Shakespeare's brain. Similarly Roger Sperry agonized over the "unthinkable thought" that mind could be more than a machine. If natural law determines the future from the present, there seems to be no place for the freedom of choice that is perceived as an element of consciousness. Quantum theory, with its probabilistic interpretation of the meaning of the wave function, appears to offer a way out of this dilemma.

But the dilemma arises because many underestimate the degree of complexity that can emerge in classical dynamics. Although the idea that classical dynamics leads inexorably to Laplacian determinism is an old one, that does not make it true. As we have seen throughout this book, the assumption that classical theory implies determinism ignores important aspects of the scientific hierarchy. The dynamics at each level of this hierarchy proceeds independently of those at lower levels; thus

$$\text{classical theory} \not\Rightarrow \text{determinism}$$

(where "$\not\Rightarrow$" is the mathematician's symbol for the expression "does not imply"). This point is important enough to merit a few more words.

As recent developments in applied mathematics have demonstrated, there are many simple dynamic systems—an ordinary differential equation of third order, for example—in which the unfolding of the present into the future can be computed only for a limited period of time. This is because computational errors grow geometrically with time (like dandelions in May) while the system is regularly forced to make critical decisions about whether to go, say, east or west from a watershed. Any such computation is based on knowledge of the initial conditions, but these are only imperfectly known because of errors in measurement and ultimately the restrictions of Heisenberg's uncertainty principle. (This problem is even more severe in biological systems because, as Bohr emphasized, a complete and exact measurement of all positions and velocities of atoms and molecules would kill the organism being studied [Bohr, 1933; Elsasser, 1969b].) Thus the size of the computer needed to predict the future after a certain interval of time grows geometrically with that interval, and it is easy to find a computation for which the computer would need to have an immense number of elements and therefore be impossible to construct.

But we have seen that the human brain is not described by a simple, ordinary differential equation. As the previous chapters have demonstrated, each level of the scientific hierarchy is defined on an immense set of possible atomistic entities for that level, and its dynamics is largely uncoupled from adjacent levels (although time and space averages over lower levels may appear as parameters in the equations describing higher levels). As one climbs to higher levels of the brain, it is no longer clear that a description in the context of differential equations is even possible. What are the relevant dependent variables at the levels of psychology or of ethnology? Do these variables have quantitative measures? How does one quantify the quality and intensity of a perception like love or hate, ecstasy or despair? What is the nature of time at these levels? Can *any* meaningful mathematical description be constructed?

When a human brain is described—with outrageous oversimplification—as a collection of McCulloch-Pitts threshold circuits, it has some hundred trillion degrees of freedom. Each millisecond is littered with watersheds; a deterministic calculation over such a short time interval is for practical purposes impossible. Yet one often hears that such a calculation is possible in principle. This claim is consonant with the collective intuition of Western science, but no one has shown it to be correct. Determinism is not a fact of science, but an assumption of certain theories and a belief of many scientists.

It is not unreasonable to view the dynamics of the mind from a broader perspective than that of systems of ordinary differential equations. Following its own rules and regularities, whatever those might be, the mind is expected to take account of ongoing cultural activities and events in ways that could alter many of the myriad watershed decisions that the brain makes in every millisecond. Although some might claim that Popper's World 3 could also be brought into this celestial computation, it is difficult to see how this might be done. There is ample scope for uncertainty and indeterminacy in mental dynamics without invoking quantum theory.

But all this is not to claim that quantum effects play no functional role in consciousness. The phenomena of lasers, superfluidity, and superconductivity demonstrate that quantum behavior can make itself known at macroscopic levels. One should note, however, that satisfactory macroscopic models of these phenomena, such as the laser rate equations and the Ginzburg-Landau equations of superfluidity and superconductivity, are classical in nature. It may be that quantum behavior of the cytoskeleton (as proposed by Hameroff and Penrose) or the synapse (as proposed by Stapp and by Beck and Eccles) influences neural computation. If this is so, the opportunities for happenstance in mental behavior would only increase.

Whether it is necessary to use quantum theory to describe mental dynamics must be decided in the laboratory of an electrophysiologist and not in the den of a theoretical physicist. In making this decision, the term "necessary" is key. It is not enough to show that a quantum mechanical calculation *could* be used to describe some phenomenon. One might, for example, choose to use quantum theory to describe the trajectory of a baseball as it speeds toward home plate at 90 miles per hour, but calculations based on Newton's second law would be no less accurate and much more convenient; thus a quantum description is not *necessary* to describe the phenomenon. Similarly, in the study of a brain, one would be required to show that a quantum interpretation is needed to understand some qualitative aspect of mental dynamics and not merely possible in principle. It is of no interest to a psychologist that a physicist chooses to make his Heisenberg cut at the highest level of brain activity unless this choice makes a difference that can be observed in a psychological experiment.

Interestingly, just such an experiment has recently been designed by psychiatrist Chris Nunn and his colleagues (1994) at a London hospital in an effort to test the theories of Roger Penrose. The basic idea of this experiment was to see whether mere *measurement* of the electroencephalogram (EEG), or brain waves, would interfere with the performance of a perceptive task. Such an interference might be expected if the measurement caused collapse of a mental wave function. The task of the subject was to identify certain numbers by pressing a button with the right thumb, and—among other results—it was observed that by altering the positions of EEG electrodes on the left side of the subject's head, the relative numbers of correct and incorrect responses were reversed. As the authors emphasize, these results are very preliminary and much work remains to be done to insure that the statistical interpretations are sound and that every experimental artifact has been eliminated. Only after all such precautions have been taken and the data are secure, will it be appropriate to consider the more fundamental question: Is the observed effect quantum mechanical or classical in nature?

CONSERVATION OF ENERGY IN THE MIND?

Karl Popper (Popper and Eccles, 1977) and John Eccles (1994) have suggested that the classical action of World 3 or World 2 dynamics on World 1 is improbable because it would violate the law of energy conservation. I disagree for several reasons.

First of all, one can readily observe such effects. Consider, for example, a human culture in which a particular ceremony involves the consumption of a mind-altering drug such as tobacco in a Zuñi kiva, alcohol at a Danish dinner party, or *ayahuasca* in an Amazonian village. In such cases, an element of World 3 reality, acting through the World 2 of a particular individual, causes the introduction of a psychologically active chemical substance into that person's body. This chemical acts on the enzymes of certain synapses, which are in World 1. Throughout the ceremony—lasting a few hours or a week—there is no evidence that the law of energy conservation, or any other fundamental law of physics, is violated. Clearly these events occur and a theorist of the mind must take account of them. Such facts must not be ignored or distorted in some Procrustean manner.

Let us next consider the neurological details. At the hierarchical level of a neuron—as we saw in Chapters 4 and 5—there are two primary dynamic activities: the all-or-nothing propagation of a nerve impulse along an active fiber, and the diffusion of transmitter molecules across a synaptic cleft. *Neither of these processes is governed by conservation of energy.* The first is an example of nonlinear diffusion in which power is balanced while energy is *not* conserved. The second process is an example of linear diffusion, which evolves like a puff of smoke on a still morning or a pseudomorph of squid ink in the ocean. Again energy is *not* conserved by the dynamics of interest. It is true, of course, that *total* energy is conserved under the dynamics of a nerve impulse. The electrostatic energy that is stored in a nerve membrane and released at the onset of sodium ion influx (near the leading edge of a nerve impulse) is ultimately dissipated by ohmic resistance to the ionic currents that flow along the axon. This loss or dissipation of energy causes a slight increase in temperature in the surrounding tissue, which is quickly offset by homeostatic regulation in warm-blooded animals. Such small temperature variations have little influence on nerve impulse dynamics and are neglected in realistic neural computations. Energy conservation is not violated by nonlinear diffusion on a nerve axon; it is just ignored.

Finally, let us consider the question from a general theoretical perspective. Conservation of energy in a dynamic system indicates that the equations governing that system do not change with the value or direction of time. In classical Newtonian dynamics, the equations of motion are identical whether time moves forward or backward, and energy conservation is a logical consequence of this fact. For diffusion processes, be they linear or nonlinear, time has an *arrow*. It proceeds from the past, through the present, and into the future. A squid's pseudomorph will disperse, but ink molecules in the ocean never reassemble themselves in a volume the size and shape of a squid. It is a logical consequence of this fact that energy is *not* conserved under the dynamics of diffusion (Fraser, 1982).

At the level of complex cell assemblies and the phase sequence in the brain, the situation is much more complicated, as this book takes pains to show. We have no clear knowledge of the temporal nature of the dynamic system, but introspection suggests that this temporality is even more peculiar than at the level of the axon and the synapse. Compare the psychological perception of 10 minutes by Isaac Newton's clock when one is reading a good book or having a chat with a friend with the same 10 minutes spent in the dentist's chair. Not only is mental time directed from the past into the future, it expands and contracts and jumps about in response to various subjective aspects of consciousness. Benjamin Libet's (1995) careful studies of the timings of neuronal activities related to sensory and volitional activities provide ample support for such informal observations. Since energy conservation requires invariance of the dynamics with respect to changes of the time variable, we must again recognize that the dynamic behavior of a brain is not constrained by energy conservation.

One of the properties of a brain is that *nothing* of dynamic importance is conserved. The analyst has no convenient constraints to impose on the brain's behavior. It is safe to conclude that energy conservation at the level of classical Newtonian mechanics provides no basis for Eccles (1994) to dismiss the emergence of mind from classical brain dynamics as mere "promissory materialism."

MATERIALISM

In the realm of science, one's attitude toward what Karl Popper called "the great tradition of materialism" is often used as an index of respectability. Those who turn away from this tradition to consider the nature of consciousness run the risk of being marked as flakes who might also believe in psychokinesis (spoon bending), mental telepathy, clairvoyance, precognition, and the like. The safest course—especially for the young scientist—is to shun such temptations and concentrate on the data from a particular level of the hierarchy. Although prudent, this approach to the gathering of knowledge ignores a pillar of the scientific tradition: face the facts.

From another point of view, the question is related to the weight one gives to emergent phenomena, or the perspective that one takes with respect to the struggle between proponents of reductionism and those of constructionism (Schweber, 1993). The subject of emergent phenomena, of course, is by no means a new one (Ablowitz, 1939), and many assert that "there is nothing new under the sun." But years spent with the study of solitons (Scott et al., 1973) and nerve pulses (Scott, 1975) have convinced me otherwise. New entities are emerging everywhere. Just as a molecule of water emerges from a mixture of hydrogen and oxygen or a protein like myosin emerges from a mixture of amino acids, the first forms of life emerged from a chemical soup some three billion years ago and new species have been emerging ever since. Because the number of possibilities is immense at each level, there will always be new molecules and proteins and forms of life. Clearly, emergent phenomena are of fundamental importance in all branches of science.

As a natural scientist, I agree that all is constructed from the particles and fields of physics, perhaps including some yet to be discovered, but that is only the preface to the story: the paper and ink on which the book of knowledge is written. This is materialism. The problem—as I see it—is with *reductive* materialism: the view that the motions of the atoms tell the *whole* story. The meaning of the message, as Roger Sperry noted, is not in the chemistry of the ink; it emerges, instead, at higher levels of reality.

FUNCTIONALISM

In an "Appendix for philosophers" Daniel Dennett (1991) admits to being to some extent a functionalist, but what does functionalism mean? Since he has proposed that a *strong* versus *weak* terminology might be useful in discussions of zombies, let us consider the following definitions.

Weak functionalism is the position that two dynamic systems are identical if their external behaviors are the same over a limited period of time. But this says little about their internal behaviors. One system could have consciousness while the other does not since—as every sophomore in electrical engineering knows—two "black boxes", as they like to call them, with identical terminal properties over a limited time can be quite different inside.

Strong functionalism, on the other hand, supposes that the dynamic behaviors of the two systems can be put into an exact one-to-one correspondence. In a silicon chip representation of a mind, this would mean that appropriate computer circuits must be arranged to mimic the dynamics of a conscious human from the individual molecules up, modeling every protein, molecule of ATP, patch of membrane, branch of the cytoskeleton, ramification of dendrite and axon, synapse, neuron, neural assembly, assembly of assemblies, and so on up to the entire cerebrum, spinal cord, musculature, and on to the relevant aspects of the cultural configuration. Assuming this could be done—which is unlikely—strong functionalism would seem to be a tautology.

DUALISM

From the perspective of one who understands the significance of emergent phenomena—I believe—it is not necessary to choose between materialism and dualism; both can be accepted, albeit with certain reservations. The concepts of World 2 and World 3 developed by Popper and Eccles seem mystical and unscientific to many physical scientists because they don't fully appreciate the hierarchical nature of nonlinear dynamic activity in a living organism.

Chapters 2 through 4 were included in this book to show in detail just how far removed are the Hodgkin-Huxley equations of nerve conduction from Schrödinger's equation of atomic physics, a view that is reinforced by the mathematical discussions in Appendices B through G. Both are excellent theories, yet they are unrelated. Even the nature of time is different in the two theories; in the Hodgkin-Huxley system there is a well-defined past and future, which in Schrödinger's equation there is not (Fraser, 1982). When one considers that there are at least as many hierarchical levels between the nerve fiber and the mind as between atomic physics and the nerve fiber (and probably more), it should come as no surprise to find that "brain-stuff" and therefore "mind-stuff" are different from "nerve-stuff." Is it dualism to recognize this? If so, then I am a dualist.

My position is close to the naturalistic dualism of David Chalmers, which was described at the end of Chapter 7. (I don't like labels, but if forced to choose, I would prefer to call it *hierarchical* or *emergent dualism.*) From both perspectives, it is assumed that consciousness arises from physical systems but in a nonreductive manner, so it is not necessary to explain in purely physical terms how conscious experience enters the picture. This nonreductiveness is the "extra ingredient" that Chalmers sees as being needed to escape the trap of mechanistic theories that purport to explain consciousness.

"But," the reader will surely ask, "what is the *nature* of this consciousness? Where does it *reside?*"

THE LOCUS OF CONSCIOUSNESS

In the hierarchical scheme presented in the introduction to this book, consciousness was rather casually placed between human culture and the brain, but this says nothing about its physical location. Where should we look for it? From introspection, consciousness seems to me a little spotlight located somewhere behind the forehead that plays over the activities of the central nervous system, seeing a new relationship between hitherto disjoint ideas in one moment and feeling a sore knee or a twinge of guilt the next. Dennett tells us, however, that we should ignore such naive impressions. The "Cartesian Theater" is an illusion, he says, because cognition is distributed about the brain. Okay, but *how* is it distributed?

As was suggested in Chapter 3, some insight into the nature and locus of consciousness can be gleaned from a consideration of the phenomenon of life. In the context of a single-celled animal, life seems to emerge from the following hierarchy of scientific levels (Goodsell, 1993):

> Cytology
> Biochemistry
> Chemistry.

The way that life might emerge from the biological hierarchy was suggested about 15 years ago by Manfred Eigen and Peter Schuster in a book entitled *The hypercycle: A principle of natural self-organization* (1979). Their basic concept is of an interrelated hierarchy of cyclic reaction networks, which they call a *hypercycle*. Recalling the discussion in Chapter 3, a hypercycle is a cycle of cycles of cycles. An example of a basic cycle is the citric acid cycle (which uses a molecule of oxaloacetate over and over again to extract the energy from acetic acid), but there are hundreds of such cycles in a simple organism. At the next level of the hierarchy, several catalytic cycles are organized into an autocatalytic cycle, which regulates its own reproduction. An example of this is the replication mechanism of single-stranded RNA. Several autocatalytic cycles are then organized at a higher level into a *catalytic hypercycle*, like a virus, which can evolve into more efficient structures. Thus life has the hierarchical form:

> Catalytic hypercycle
> ↑ ↓
> Autocatalytic cycles
> ↑ ↓
> Basic catalytic cycles.

In these nested and interacting cycles, energy, mass, and information are all currencies that are traded back and forth along the arrows. Biophysicist Ronald Fox (1982) relates such cycles to the *uroboros*, a mythical monster that eats its own tail, in a fascinating book that describes many of the biochemical details.

The importance of such multi-level dynamics is often ignored by theorists. In a compelling survey of contemporary biological theory, Brian Goodwin (1994) has recently suggested that biology has become too "genocentric" over the past century and should return to its "organocentric" roots. Basing his arguments on recent developments in nonlinear mathematics—and nonlinear diffusion, in particular—he shows that living organisms display emergent properties that are not embodied in their genetic codes. Without denying the importance of molecular biology, Goodwin takes the position that organisms are not mere molecular machines, but "functional and structural unities" that are "as real, as fundamental, and as irreducible as the molecules out of which they are made."

Biomathematicians usually concentrate their attention on a single level of the scientific hierarchy, but the understanding of hypercycles requires a broader perspective. Instead of a description that is expressed as a single system of differential equations, a nested hierarchy of such mathematical descriptions is needed. Averages over the variables at one level may appear as parameters in the higher-level systems, and instantaneous values of the variables at a higher level establish the structures that are seen at lower levels. A number of biologists and applied mathematicians are beginning to consider life from the perspective of Eigen and Schuster (Baas, 1994; Farmer et al., 1986; Nuño et al., 1993), and it is my belief that studies of the dynamics of such hierarchical structures will become increasingly important in the years ahead.

In one of the more striking of these efforts, Walter Fontana and Leo Buss (1994) have recently developed a minimal theory of biological organization that generates higher-order structures from a simple representation of molecular interactions. This algebraic theory, called the *lambda-calculus*, was introduced in 1941 by Alonzo Church as a means to study the computability of mathematical algorithms. (In the briefest possible terms, it can be said that Church's lambda-calculus puts mathematical functions and their arguments on an equal theoretical basis.) Fontana and Buss show that the lambda-calculus provides a natural setting for considering a question that arose in Chapter 3: How do the atoms organize themselves into molecular structures? In the context of this calculus, they find that the atoms in (the mathematical equivalent of) a well-stirred pot organize themselves into different hierarchical levels, where "Level 0 is defined by self copying objects or simple ensembles of copying objects. Level 1 denotes a new object class, whose objects are self-maintaining organizations made of Level 0 objects, and Level 2 is defined by self-maintaining metaorganizations composed of Level 1 organizations." The corresponding diagram:

Level 2
↑ ↓
Level 1
↑ ↓
Level 0

is closely related to that of Eigen and Schuster for the structure of life.

Note carefully how the mathematics is being used here. It is not used to study the dynamic details of a life process but rather to describe the way that a creature becomes organized. Might a similar approach be applied to the study of consciousness?

From the previous discussions, an appropriate hierarchy for the dynamics of the mind is:

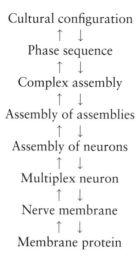

Cultural configuration
↑ ↓
Phase sequence
↑ ↓
Complex assembly
↑ ↓
Assembly of assemblies
↑ ↓
Assembly of neurons
↑ ↓
Multiplex neuron
↑ ↓
Nerve membrane
↑ ↓
Membrane protein

where the currency passed back and forth along the arrows is now information alone. Energy and mass are mere housekeeping details, no more important than yesterday's trash.

At the level of the nerve membrane, consciousness can be altered by chemicals that bind to membrane proteins and block ionic transport across the membrane. At the level of the neuron, consciousness is routinely switched off and on by anesthetic agents that change the actions of the synaptic contacts between cells. At higher levels, one is conscious of something called thought, which is stored in the myriad complex assemblies that have been pieced together throughout the years of learning. Thought, in turn, is formed by and interacts with the culture in which it develops. Up and down the hierarchy, from membrane ion channels to the ebb and flow of cultural interactions, an intense intercourse between the levels continues.

Just as life emerges from cycles of cycles of cycles of biochemical activity, consciousness seems to emerge from assemblies of assemblies of... of assemblies of neurons. As we have noted in Chapter 7, this is what Erich Harth (1993) calls a "creative loop," and it is also called a hyperstructure by the Norwegian mathematician Nils Baas (1995). Some might argue that a hierarchical structure in the mind is merely a *mechanism*, and being a mechanism, it cannot *explain* consciousness; it cannot provide a solution to what David Chalmers calls the "hard problem." What, then, is the essence of consciousness?

An answer to this question requires the specification of an "extra ingredient" beyond mere mechanism. Traditionally this ingredient has been called the *soul*, although the behaviorists dealt with the hard problem by denying it. From the perspective of natural science, both of these approaches are unacceptable. And what is left? As we have seen in Chapter 7, three suggestions have been put forward to explain the nature of the mind: First, it has been proposed that consciousness may be embodied in a quantum mechanical wave function. Second, some feel that it may be a new *primitive*, a fundamental property like the mass or electrical charge of an elementary particle. Finally, it is suggested here that consciousness may emerge from several levels of the mental hyperstructure in a nonreductive manner. This is in accord with the views of Goodwin (1994) who observes:

> It is another of those curious paradoxes that a large number of scientists who work in the area of artificial intelligence, and the cognitive sciences generally, deny that consciousness has any fundamental reality and say it is basically an epiphenomenon of brain activity—the electrical and molecular processes that go on in brain cells. This is just like the denial on the part of many biologists that organisms have any fundamental reality that cannot be explained by genes and molecular activities.

If the hierarchical emergence of consciousness makes sense, it follows that research into its nature must be truly interdisciplinary. In the study of life it is generally recognized that the joint efforts of chemists, biologists, biophysicists, biochemists, cytologists, physiologists, and physicians are all required in order to make progress. Similarly, research on consciousness should meld the activities and insights of biochemists, cytologists, electrophysiologists, neuroscientists, engineers, computer scientists, physicians (particularly anesthesiologists, neurologists, and psychiatrists), psychologists, sociolologists, and ethnologists. Philosophers, physicists, and mathematicians may also have modest roles to play, but they should not seek to dominate the discussion. It is not necessary that every research team must include all these professions, but institutes that are devoted to studies of consciousness should strongly encourage such interactions.

Interdisciplinary research is often difficult to arrange because scientists have strong tribal instincts and tend to work on "real biochemistry," "real electrophysiology," "real psychology," or "real ethnology." This is of course entirely understandable. It takes much talent and years of hard work to become a first-rate biochemist, electrophysiologist, psychologist, or ethnologist. But to understand consciousness we must get beyond these parochial limits. We must set aside the outdated assumption that it is somehow more scientific to concentrate on a single layer of the hierarchy. The really exciting science nowadays is interdisciplinary. The new *Journal of Consciousness Studies* seems to be a step in this direction. The editors of this journal aim to

> provoke serious, spirited debate by actively seeking opposing views, and are prepared to include a wide diversity of topics and approaches, from hard science to spiritual metaphysics; from the cultural imperialism of the 'information superhighway' to deconstructionism.

It will be fun to watch this debate unfold.

ARTIFICIAL INTELLIGENCE

The developments of computing technology over the past half century have been impressive. Following the invention of the transistor in 1950 and the development of integrated circuits in the 1960s, the "giant brains" of the 1950s have shrunk to gadgets the size of a thumbnail, and the computing power of desk-top machines grows year by year. It is quite understandable that the computer scientists and engineers who design this equipment should be optimistic about the possibility of creating machines that think, but it is important for the rest of us to be realistic. What do we mean when we say that a machine is "thinking"?

One answer—proposed by Alan Turing in 1950 and described in Chapter 7—may be too easy. The mark of an intelligent system is not merely to convince someone through conversation in a limited period of time that it or he or she is intelligent; political candidates do this all the time. The task is to *do* something intelligent: translate a song, construct an imaginative mathematical theorem, argue effectively, whatever—and such activities require World 3 interactions.

Pursuing this line of thought, we recognize that mental dynamics, as one normally considers the term, involve interactions with human culture, and consciousness emerges out of these interactions as both the apex of the most complex dynamic object in the universe (the human brain) and an atom of human culture. At each moment our conscious minds ride the turbulent waves of a hundred million past perceptions and impressions even as we attempt to interact—via touching, looking, talking, singing of songs, reading and writing of films, books, poems, newspapers, television, fax, e-mail, and all—with millions of others.

A truly intelligent machine, it seems to me, would require a means to interact with a culture, and it would also need to be conscious. And computer consciousness will not miraculously appear on the day that some machine gets good at the Turing test, as some functionalists seem to suggest. It would only be a feature of a machine if a computer engineer were to design it in. The wings of an airplane, after all, do not start to flap because it passes a flying test. From this perspective, the philosophy of functionalism may be an impediment to the development of computer science.

It should be emphasized that the human brain operates at a level of subtlety and complexity that lies far, far beyond anything that is imagined by present-day computer engineering. We should not be misled by the stale hyperbole that is printed year after year in the Sunday papers; true artificial intelligence will not be with us soon. Nonetheless, I see no way to prove that machine intelligence is impossible. How can one put limits on the technical prowess of future engineers?

But a truly intelligent machine must, in Erich Harth's terms, "be its own client"—like HAL in the film *2001*, who responds to a human command to switch itself off with the chill phrase:

I'm sorry, Dave, I'm afraid I can't do that.

Or Mary Shelley's (1818) creature, who, at the end of her tale, laments:

> I have destroyed my creator, the select specimen of all that is worthy of love and admiration among men... There he lies, white and cold in death.

Since a truly intelligent machine might be alien to human concerns, we may find that we don't *want* it.

TWO DREAMS

Finally, allow me to mention two vivid dreams I had many years ago, while immersed in studies of physical science. In the first, I was wandering alone on a barren, lunar landscape where the rocks were composed of jagged crystals, their atoms smugly arranged in regular patterns. During this dream, I experienced an intense feeling of *horror*: there was nothing in this world but the laws of physics and chemistry. Life, complexity, beauty did not exist. All was clear and logical and computable but without emotion. A few nights later, in another dream, a living rose was illuminated by the warm light of a setting sun. In contrast to the previous dream, I felt a sublime *joy* in being part of a world of such splendor.

Experiences of this sort—I admit—color my attitudes toward the various discussions of the philosophy of consciousness surveyed in the previous chapters. This is to be expected. Philosophy, as opposed to science or mathematics, is less an exercise in reasoned thought than a rationalization of ideas that we are predisposed to believe. If it were based entirely on reason, William James commented somewhere, all philosophers would agree.

My own contribution, I would hope, is to impress upon the reader the importance of *not* choosing between the worlds of the first and second dreams. Somewhere between the rational realm of the first dream, where science reigns, and the enchanted realm of the second dream, where life's mysteries still lie, there is a viable intermediate world. There, one day, it will be possible for science to understand the nature of consciousness without denying or neglecting its more intangible aspects.

References

R Ablowitz. The theory of emergence. *Philos. Sci.*, 6:1–16, 1939.

H Adams. *Mont Saint Michel and Chartres*. Penguin Classics, New York, 1986 (First published in 1904).

P W Anderson. More is different: Broken symmetry and the nature of the hierarchical structure of science. *Science*, 177:393–396, 1972.

N A Baas. Emergence, hierarchies, and hyperstructures. *Artificial life III*, C G Langton, editor. Addison-Wesley, Reading, Massachusetts, 1994.

N A Baas. A framework for higher order cognition and consciousness. In *Toward a science of consciousness*, S R Hameroff, A W Kaszniak, and A C Scott, editors. MIT Press, Cambridge, Massachusetts, 1995.

N Bohr. Light and life. *Nature*, 131:421–423, 457–459, 1933.

A Church. *The calculi of lambda-conversion.* Princeton University Press, Princeton, New Jersey, 1941.

D C Dennett. *Consciousness explained.* Little, Brown, Boston, 1991.

J C Eccles. *How the self controls its brain.* Springer-Verlag, Berlin, 1994.

M Eigen and P Schuster. *The hypercycle: A principle of natural self-organization.* Springer-Verlag, Berlin, 1979.

W M Elsasser. Acausal phenomena in physics and biology: A case for reconstruction. *Amer. Sci.*, 4:502–516, 1969a.

W M Elsasser. The mathematical expression of generalized complementarity. *J. Theoret. Biol.*, 25:276–296, 1969b.

J D Farmer, S A Kauffman, and N H Packard. Autocatalytic replication of polymers. *Physica D*, 22:50–67, 1986.

W Fontana and L W Buss. "The arrival of the fittest": Toward a theory of biological organization. *Bull. Math. Biol.*, 56:1– 64, 1994.

R F Fox. *Biological energy transduction: The Uroboros.* John Wiley & Sons, New York, 1982.

J T Fraser. *The genesis and evolution of time.* Harvester Press, Sussex, 1982.

M Gell-Mann. *The quark and the jaguar.* W. H. Freeman and Company, New York, 1994.

D S Goodsell. *The machinery of life.* Springer-Verlag, New York, 1993.

B Goodwin. *How the leopard changed its spots: The evolution of complexity.* Charles Scribner's Sons, New York, 1994.

E Harth. *The creative loop: How the brain makes a mind.* Addison-Wesley, Reading, Massachusetts, 1993.

E Harth. Self-referent mechanisms as the neuronal basis of consciousness. In *Toward a science of consciousness*, S R Hameroff, A W Kaszniak, and A C Scott, editors. MIT Press, Cambridge, Massachusetts, 1995.

B Libet. Neural time factors in conscious and unconscious mental functions. In *Toward a science of consciousness*, S R Hameroff, A W Kaszniak, and A C Scott, editors. MIT Press, Cambridge, Massachusetts, 1995.

C M H Nunn, C G S Clarke, and B H Blott. Collapse of a quantum field may affect brain function. *J. Consciousness Stud.*, 1:127–139, 1994.

J C Nuño, M A Andrade, F Morán, and F Montero. A model of an autocatalytic network formed by error-prone self-replicative species. *Bull. Math. Biol.*, 55:385–415, 1993.

K R Popper and J C Eccles. *The self and its brain.* Springer-Verlag, Berlin, 1977.

S S Schweber. Physics, community and the crisis in physical theory. *Physics Today*, 46:34–40, November 1993.

A C Scott, F Y F Chu, and D W McLaughlin. The soliton: A new concept in applied science. *Proc. IEEE*, 61:1443–1483, 1973.

A C Scott. The electrophysics of a nerve fiber. *Rev. Mod. Phys.*, 47:487–533, 1975.

M W Shelley. *Frankenstein (or the modern Prometheus)*. Dell, New York, 1965 (first published in 1818).

R W Sperry. Forebrain commissurotomy and conscious awareness. *J. Medicine and Philosophy*, 2:101–126, 1977.

L Weiskrantz. *Blindsight: A case study and implications*. Oxford University Press, Oxford, 1986.

"Then I Tell Them
What I Told Them"

Queen:
This is the very coinage of your brain.
This bodiless creation ecstasy
is very cunning in.
Hamlet:
Ecstasy?

William Shakespeare

T he story is often related of a successful preacher who was asked to reveal the secret of delivering a good sermon. "First," he replied, "I tell them what I'm going to tell them. Then I tell them. Then I tell them what I told them."

In this spirit it seems appropriate to look back and consider where we have been; thus this chapter reviews the salient observations of Chapters 2 through 9. While leafing through these final pages, the reader should pay particular attention to the diminishing value of mathematical description at upper levels of the hierarchy. As science becomes more biological, neurological, psychological, or ethnological, it becomes progressively less analytical, and controlled experiments become ever more difficult to prepare.

Strangely, this lack of precision has been a source of embarrassment for the social sciences. It should not be. Paradoxically, physics is called a "hard" science because it restricts itself to the easy problems for which currently available mathematical tools are effective. Psychology, on the other hand, is called a "soft" science because it deals with reality on a level of complexity at which problems are difficult because sharp analytical tools have not yet been—and may never be—developed. It is the physicist, rather than the psychologist, who should feel uncomfortable about this division of labor.

RECAPITULATION
Atomic Physics and Chemistry

We began with atomic physics—the realm of quantum mechanics—and we saw how this strange theory, with its mathematical operators and eigenfunctions and collapsing wave packets and conscious observers and probability amplitudes and indeterminacy principles, can predict the spectrum of the

179

hydrogen atom with impressive precision. With certain assumptions and approximations, Erwin Schrödinger's famous equation reveals the structures of many electron atoms in the periodic table of the elements and much about the nature of chemical bonding. When we asked how atoms interact, however, we found that the quantum theory was too difficult to unravel and it was necessary to turn to classical theory and experimental measurements to estimate the potential energy surface on which they move.

In this transition from atomic physics to chemistry we met our first riser—or *combinatorial barrier*—on the stairway to the mind. Using a term coined by Walter Elsasser, we found the number of molecules that can be constructed from one hundred odd atoms to be *immense* because it is greater than

$$\Im = 10^{110},$$

a number that equals the atomic mass of the universe (the mass of the universe divided by the mass of a hydrogen atom) multiplied by its age measured in picoseconds. Since there is no way that all possible molecules could be investigated, chemists will never want for new and interesting problems to consider.

The dynamic aspects of chemistry—the Newtonian equations of molecular dynamics and the rate equations for chemical combination—are almost entirely divorced from quantum theory. They exist as autonomous levels in the scientific hierarchy, and they differ from quantum theory in a fundamental way. Quantum mechanics is a *linear* theory. The initial conditions for a particular problem can be divided into convenient components and the dynamics of each component can be computed independently of the others. For this reason the solution to a quantum problem can be expressed as a *wave packet*, but it is incorrect to assume that the same analytical structure holds at higher levels of the hierarchy. The equations of molecular dynamics and the chemical rate equations, for example, are *nonlinear*. Each problem must be treated as an individual case, and it becomes difficult—if not impossible—to map out all possible behaviors. Above atomic physics, all levels of the scientific hierarchy are nonlinear.

Although physicists have made a reasonable case that "all of chemistry" (in Paul Dirac's words) can be derived "in principle" from the properties of electrons and atomic nuclei using Schrödinger's equation, it is not an airtight case because they have not been able to do it. The disappointing results of quantum chemistry demonstrate the difficulties that arise when one attempts to penetrate the easiest of combinatorial barriers.

Biochemistry

With biochemistry we met the first truly complicated level of science. Situated at the verge of viability, the chemistry of life deals with molecules of amino acid, ribose, deoxyribose, cytosine, uracil, adenine, guanine, thymine, lipid, etc., which engage in extremely complex, nonlinear interactions to form biomolecules. These biomolecules, then, are the atomistic entities upon which life is based.

As an example of this complexity, we considered myoglobin, one of the simpler proteins, which has been called the "hydrogen atom of biochemistry." Composed of 2318 atoms with the chemical formula

$$C_{738}H_{1166}FeN_{203}O_{208}S_2 \, ,$$

the structure of myoglobin (see Figure 6) can be understood as a molecular chain of amino acids, like a string of colored beads. The motion of myoglobin is even more complex than its structure. Described as "screaming and kicking" on the basis of a wide variety of experiments, this motion evolves from a hierarchy of conformational substrates.

Since our understanding of the dynamics of biomolecules is based on the concept of a classical potential energy function, it is just at this level of approximation that classical nonlinearity strides in the front door and the linear quantum theory slips out the back window. Biomolecular dynamics is almost entirely a classical science. In considering whether life is based on the laws of physics, we noted Schrödinger's opinion that it is not. "And that is not on the ground that there is any 'new force' or what not, directing the behaviour of the single atoms within a living organism," as he says, "but because the construction is different from anything we have yet tested in the physical laboratory."

The Nerve Axon

Our next step up the hierarchy of knowledge was a large one. We considered the nerve axon, a cylindrical tube bounded by a lipid bilayer membrane. Imbedded within this membrane are *intrinsic membrane proteins*, the atomic structures of which are not known because lipid-soluble proteins are yet to be crystallized for x-ray diffraction studies. Since they mediate the transmembrane flow of potassium and sodium ions, these proteins are key elements in the electrodynamics of a nerve membrane, but their electrochemical properties are known only from experimental measurements. Theory tells us little.

Based upon Kenneth Cole's measurements of ionic currents through the nerve membrane, Alan Hodgkin and Andrew Huxley in 1952 developed a set of nonlinear partial differential equations that correctly predict the time course of a nerve impulse, as shown in Figure 10. It is interesting to note that these H-H equations are similar to equations that were introduced in the same year by Alan Turing as a basis for the generation of biological form (morphogenesis). The technical name for the processes described by such equations is *nonlinear diffusion*.

Without going into the mathematical details, the process of nonlinear diffusion supports the emergence of traveling waves that release energy stored in the medium, and this energy, in turn, supplies the power that is dissipated in the wave. Thus the wave travels at a speed that is equal to the power dissipated divided by the energy stored per unit length of the system because at this speed the wave *digests* energy at the same rate that it *eats* it. This simple concept describes the propagation of an *all-or-nothing* impulse along the axon of a nerve cell, and the *threshold* for stimulation of that pulse.

Although Michael Faraday—the brilliant nineteenth-century experimental physicist and physical chemist—had emphasized the candle's importance as a subject for philosophical speculation, it was not until the late 1960s that the applied mathematics community in western Europe and America began to take the subject seriously (although flame propagation had been studied a decade earlier in the Soviet Union). The candle flame is now recognized as a metaphor for a nerve impulse: the "travelling point of light" that Charles Sherrington used to represent an atom of thought.

It is important to note that the nonlinear diffusion equations—which were developed by Hodgkin and Huxley to describe the dynamics of nerve—*are independent of the equations of physics and chemistry*. The H-H equations don't violate the laws of physics and chemistry, but they cannot be derived from them. They are new laws appropriate to the science of electrophysiology, which is removed by several hierarchical levels from atomic physics.

The Neuron

At the level of the nerve cell, we compared two models of a real neuron:

The McCulloch-Pitts neuron. Introduced by Warren McCulloch and Walter Pitts in 1943, this model represents the neuron as a simple device that compares a weighted sum of input signals with a threshold level. If the weighted sum of the inputs exceeds the threshold, the neuron *fires* and transmits all-or-nothing pulses to its output terminals. Although it is the workhorse of much current research in neural network theory, this model vastly underestimates the computing power of a single neuron.

The multiplex neuron. Based upon theoretical and experimental research during the 1960s and 1970s, it is now known that a neuron is a much more sophisticated information processor than a McCulloch-Pitts neuron. Since an individual segment of nerve fiber has a threshold for firing, branching regions on the input (dendritic) side of a neuron may act as gates for Boolean addition or multiplication. Coupled with observations of high-frequency blockage on the output (axonal) side and the possibility of internal information processing in the cytoskeleton, one arrives at the concept of a *multiplex neuron* that is more like a computer chip (or integrated circuit) than a single gate (or transistor).

Recent experiments suggest that even the multiplex neuron oversimplifies the dynamic complexity of a real neuron. Electron microscopy reveals a profusion of dendrodendritic synapses, sodium and calcium ions that act as nonlinear dendritic amplifiers, and synaptic activity that can alter the information processing in a neuron on several time scales. At present it is not clear how all of these effects might be embodied in a computer model of a nerve cell.

Neural Networks and the Brain

At the level of the human brain, we began with Sherrington's image of an "enchanted loom" and asked how its observed activity could be related to the behavior of its constituent neurons. Assuming, for the sake of argument, that the neocortex is composed of ten billion McCulloch-Pitts neurons, each of which has ten thousand input connections, the number of brains that can be realized is greater than

$$\sim \Re^{10^{16}},$$

which is the immense number multiplied by itself ten thousand trillion times! This is a combinatorial barrier that is much higher than those between atomic physics and chemistry and between chemistry and biochemistry. The awesome height of this barrier—calculated under the simplest possible assumptions—suggests that several hierarchical levels of dynamic activity are to be expected between the neuron and the global properties of the brain, but how might such levels be organized?

At midcentury, when North American psychology was dominated by the conflicting claims of Gestalt theory, stimulation-response connectionism, and behaviorism, the psychologist Donald Hebb introduced the concept of a *cell-assembly*, which he defined as

> a diffuse structure comprising cells in the cortex …, capable of acting briefly as a closed system, delivering facilitation to other such systems and usually having a specific motor facilitation.

A most important feature of the cell assembly is its hierarchical nature. Each complex assembly is composed of subassemblies and can itself serve as a subassembly for other assemblies of higher order. Just as a neuron fires when a sufficient number of its inputs become active, so will an assembly ignite—like a bonfire—when a sufficient number of its subassemblies become active.

Among the variety of observations that support Hebb's theory, the most convincing are these:

Stabilized-image experiments. As shown in Figure 24, it is possible to directly observe the appearance and disappearance of subassembly components of a simple geometrical figure such as a square or a triangle.

Mapping of the brain's visual architecture. David Hubel and Torsten Wiesel investigated the functional architecture of the cat's visual cortex and found that neurons could be classified as simple, complex, and hypercomplex (and at least one of still higher order) according to the visual patterns that cause them to fire. Based on this physiological evidence, one can relate simple cells in the visual cortex to components in primary or first-order assemblies, complex cells to components in secondary assemblies, hypercomplex cells to third-order assemblies, and so on. Thus does the brain exhibit several hierarchical levels of dynamic organization between the switching of a single neuron and the experiential hallucinations induced by Wilder Penfield in the course of his operations to relieve the symptoms of epilepsy.

In comparing mammalian species, it is found that adult thinking ability increases with the ratio of association to sensory areas of the cortex. This is expected from the perspective of Hebb's cell assembly theory because a larger association cortex provides more room for the storage of latent assemblies. Based on extremely conservative assumptions, the number of high-order assemblies that the human brain can store is

$$\sim 10^9,$$

which is about equal to the number of seconds in thirty years. During childhood these assemblies are constructed day by day to form the basis for adult thought.

Consciousness

At the hierarchical level of human consciousness it is not possible to report a consensus of the scientific community because there is none. Materialists, functionalists, and dualists are—according to a recent issue of the popular science magazine *Omni* (October 1993)—engaged in

> slinging mud and hitting low like politicians arguing about tax hikes. Although the epithets are more rarefied—here it is "obscuritanist" and "crypto-Cartesian" rather than "liberal" and "right wing"—recent exchanges between neuroscientists and philosophers of mind (and in each group among themselves) feature the same sort of relentless defensiveness and stark opinionated name calling we expect from irate Congressmen or trash-talking linebackers.

To the extent that this is a true appraisal of the current status of consciousness research, it is unfortunate. Like life, the phenomenon of consciousness is intimately related to several levels of the scientific hierarchy, so the appropriate scientists—cytologists, electrophysiologists, neuroscientists, cognitive scientists, psychologists, psychiatrists, neurologists, anesthesiologists, sociologists and ethnologists—should be working together. It is difficult to see how this elusive phenomenon might otherwise be understood.

Chapter 7 presented an eclectic survey of scientific views of consciousness over the past century. Many direct quotations were included so the reader can draw his or her own conclusions. My perspectives have influenced the selections and the emphasis of this survey, but it is hoped that the chapter is representative of modern thinking about the nature of consciousness.

Although the past hundred years has been a period of intense study in this area, the key questions being considered by researchers today are almost identical to those that perplexed William James at the close of the nineteenth century.

The main thesis of this book is that many of the differences between varied philosophical attitudes toward consciousness evaporate as one begins to appreciate the importance of *emergent* phenomena in the hierarchical description of nature. In particular:

Materialism. If materialism is taken as the view that all reality can be expressed as implications of the laws of physics and chemistry, it is unrealistic. As was noted in Chapter 3, the protein folding problem hasn't yet been solved by biophysicists, just as the equations of hydrodynamics remain a mystery to applied mathematicians after two centuries of intense study. One finds new laws and regularities emerging at each level of the hierarchy of knowledge, and each set of such laws expresses the nature of scientific reality at a particular level. These new laws can be traced to the laws of physics and chemistry, but reductive materialism is not tenable. It is not possible to construct the details of biology from the facts of physics and chemistry.

Functionalism. I don't understand the concept. If functionalism means that two systems with the same external behavior (over a limited period of time, of course) are identical, then it is wrong because it is possible to construct counter-examples. If functionalism means that one system (say a computer based on silicon transistors) models *every aspect* of a human brain, then the position is both unrealistic and trivial.

Dualism. The World 2 (of inner life) and World 3 (of cultural reality) in the perspective of Karl Popper seem mystical and nonscientific to many physical scientists because they don't fully appreciate the hierarchical nature of nonlinear dynamic activity in a living organism. If one considers that the nonlinear diffusion system of Hodgkin and Huxley is as far removed from atomic physics as consciousness is from the multiplex neuron, the question appears in a different light. Recalling that conduction of a nerve impulse seemed to require an *élan vital* for its explanation in the nineteenth century, though it can be modeled by a simple candle, we might expect a corresponding explanation for consciousness to emerge when and if we understand it.

I do not share recently expressed beliefs that the mysterious nature of consciousness derives from the mysterious nature of quantum theory via some global wave function in the brain. Such theories stem from a visceral, and therefore unreflecting, acceptance of the unproven claims of reductive determinism.

Manfred Eigen and Peter Schuster have shown that *hyperstructures*—involving several layers of the scientific hierarchy—provide an appropriate basis for the phenomenon of life. Hyperstructures in the brain may play the same role with respect to the phenomenon of consciousness. In this connection, I am reminded of the often quoted admonition by the mathematician Stan Ulam:

> Don't ask what mathematics can do for biology. Ask instead what biology can do for mathematics.

The phenomenon of consciousness is difficult to comprehend, it seems to me, because it plays central roles in two levels of knowledge that are of great interest to us: consciousness is the highest level of organization in the brain, and it is an atom at the level of human culture.

Ethnology

During the first four decades of this century, the science of ethnology (or cultural anthropology) in the United States was dominated by Franz Boas, the father of American anthropology. Although trained in physics, Boas rejected the claims of reductive materialism early in his scientific career, but he did not succumb to mysticism in his attempts to understand the mysteries of social behavior. On the contrary, he was a devoted skeptic with profound respect for scientific accuracy and carefully recorded experimental data.

While deeply suspicious of unfounded theoretical speculation, he understood the importance of the two aspects of dynamics that have become widely recognized and appreciated only in the past twenty years: *nonlinearity* and *chaos*. His view of human culture as emerging from the nonlinear interactions of human minds was based on this understanding.

As Ruth Benedict's classic analysis of the Zuñi, Dobu, and Kwakiutl cultures shows, our paradigm for the emergence of form within each level of the scientific hierarchy is again appropriate. As a result of its peculiar history, each culture selects, from "a great arc" of possibilities, the elements it needs to organize its particular configuration of human life. Although the arc is defined by the genes, the choice is made by the culture, or as a Digger Indian Chief casually expressed it to Benedict one afternoon:

> In the beginning God gave to every people a cup of clay, and from this cup they drank their life.

FINAL COMMENTS

The ladder with which we began our discussion still stands against a tree in the orchard. In the warmth of an afternoon sun, we remain *aware* of the softness in its silver patina.

Having climbed our metaphorical ladder—rung by rung, from atomic physics to the configurations of human culture—what have we learned? Has this climb helped us to understand how we become aware? The candid answer to this question is negative, but we do have a clearer picture of how to conduct the search.

The idea that all can be reduced to the spare concepts of physics and chemistry has been exposed as untenable because each level of the hierarchy is dynamically independent of its neighbors. Dynamic independence—in turn—arises from *nonlinearity*, which induces the *emergence* of new and qualitatively different *atomistic entities* at each level.

Remaining respectful of the beauty of mathematical formulation, the skills of experimental science, and the power of modern computing machines, we have come to realize that the complexity of a truly interesting system—a living organism, a mind, a human culture—is awesome. We have watched the rigor of mathematics become ever less appropriate as we climbed the stairway to the mind. At higher levels, as systems become more interesting for their unique characteristics, they begin to defy a general analysis. *Happenstance* cannot be ignored or explained away with careless assertions of scientific determinism. A child, a poem, an affair of the heart—each has a life of its own, a life that springs from an immense well of historical accident and organic flexibility.

We have sought to remain within the confines of natural science but have not been restrained by those of reductive materialism. As natural scientists, we have tried to fulfill our responsibility to look outside of ourselves and *see what is there*. Theory has been used with restraint, only where it seems appropriate. At higher steps of the stairway, we have ceded the desire to predict and control, seeking merely to understand.

Have we managed to stay within the bounds of natural science? Or is "emergent dualism" beyond the pale? Have we carelessly become "crypto-animists" or mystics pretending to be naturalists? Each must answer this question for himself or herself with honesty and humility.

This is not an easy task because we physical scientists—let's face it—are not humble. We tend toward intellectual arrogance. Surrounded by our conservation laws, our mass spectrometers, and our digital computers, we often look with a mix of pity and scorn on those who struggle with the ill-defined problems of real human interest. This attitude is incorrect, and it can change. Indeed, it must change. We must learn to be appropriately respectful of those activities that we are not able to dominate. We must learn this not only for our own sakes—as fully developed human beings—but also to make possible our rightful and significant contributions to the understanding of life, mind, and human culture.

Back in the orchard the real ladder still leans against the apple tree, and we remain aware of it, softly gray on an autumn afternoon. But we do not—for all our efforts—understand our awareness. The "hard problem" remains. None of us yet can say what conscious awareness is.

If we are ever to achieve this level of understanding within the context of natural science, all of us must work together. Physicists and physiologists, neurologists and ethnologists, biochemists and philosophers, engineers and poets must abandon false pride and outworn tribal loyalties in order to learn how to learn from each other. Only in this way can our collective imagination transcend the limited perspectives of a single rung of the metaphorical ladder and approach an understanding of our awareness of the real one.

Nonlinearity and Emergence

T he term *nonlinear* is defined in the context of the relationship between cause and effect. Suppose that a series of experiments on a certain system has shown that cause C_1 gives rise to effect E_1; thus

$$C_1 \rightarrow E_1, \qquad (A.1)$$

and similarly

$$C_2 \rightarrow E_2 \qquad (A.2)$$

expresses the relationship between cause C_2 and effect E_2, where "$\rightarrow$" indicates the action of the system being studied. This system is *linear* if

$$C_1 + C_2 \rightarrow E_{12} = E_1 + E_2 \qquad (A.3)$$

and *nonlinear* if

$$C_1 + C_2 \rightarrow E_{12} \neq E_1 + E_2. \qquad (A.4)$$

Equation (A.3) indicates that for a linear system the cause can be arbitrarily divided into convenient components (C_1) and (C_2), whereupon the effect will be correspondingly divided into (E_1) and (E_2). This property is useful for analysis but rare in the biological world.

More common is the nonlinear situation shown by Equation (A.4), where the effect from the sum of two causes is not equal to the sum of the individual effects. The whole is not equal to the sum of its parts. Nonlinearity is less convenient for the analyst because more things can happen; thus it is more interesting for the supreme synthesizer: Mother Nature.

Linear systems are much easier to analyze than nonlinear systems, since a complex cause can be expressed as a convenient sum of simple components, and the combined effect is the sum of the effects from each component of the total cause. For this reason linear systems have been favored by physical scientists, especially during the current century. In the nonlinear domain of biology, however, one bite stimulates the appetite while ten satisfy and twenty nauseate; a story told once can be amusing but told over again it becomes boring if not painful; one sperm will fertilize an egg, two can do no more. Nonlinear systems are more difficult to analyze because they are more interesting. More things can happen—new atomistic building blocks may *emerge* at each hierarchical level—and that is why the realm of biology is so rich (Ablowitz, 1939).

As a simple example of the way that nonlinearity can lead to the emergence of a new dynamic entity, consider a large trampoline that is constructed from a rubber sheet. The *escape energy* of a child on the trampoline is the work that he or she must do to climb up to the edge. Suppose, to be specific, that the escape energy is equal to the square of the child's weight. Then the escape energy for two children (of weights a and b) standing together is

$$\text{Escape energy} = (a + b)^2 = a^2 + b^2 + 2ab.$$

Thus $2ab$ is the *extra* escape energy that appears because the two children are standing together on the trampoline. If they try to move apart, they must do $2ab$ units of work. As a pair, they are held together by this energy of attraction or binding energy. Thus the *pair of children* is a new atomistic entity that emerges because the trampoline is nonlinear.

A team of bicycle racers, a flock of geese, and a school of fish provide additional examples of this effect where the emergent team, flock, and school are held together by attractive energies that arise because the surrounding fluid (air or water) moves with them. But there are many variations on the theme. A molecule of benzene is bound by a combination of electrostatic and chemical valence (quantum) energies; the membrane of a living cell organizes itself as a molecular bilayer in response to the shape and charge distribution of lipid molecules and the large dielectric constant of water; and the flame of a candle expresses a balance between the nonlinear localizing effect of combustion and the dispersive effect of thermal diffusion. At higher levels one can consider a burst of nervous activity, a memory, or a thought. It is, in fact, difficult to think of a biological "thing" that does not take advantage of nonlinearity to establish its integrity, its oneness.

The Mathematics of
Quantum Theory

I n December of 1925, Schrödinger constructed the equation

$$\left[-\frac{h^2}{8\pi^2 m_e} \nabla^2 + V(x, y, z) \right] \psi = E\psi, \qquad \text{(B.1)}$$

and his "few calculations" involved showing that this equation predicts the spectrum of electromagnetic radiation (visible light, infrared radiation, etc.) from the hydrogen atom. Experimentalists had found this spectrum to occur only at particular frequencies, $f(n, m)$, given by the experimental formula

$$f(n, m) = R \left(\frac{1}{n^2} - \frac{1}{m^2} \right). \qquad \text{(B.2)}$$

The quantity R in Equation (B.2) is an experimentally determined constant (called the Rydberg energy), $n = 1, 2, 3, \ldots$, and m is an integer greater than n.

How can the relationship between Equations (B.1) and (B.2) be understood? In Equation (B.1) the symbol ∇^2 stands for the *operator*

$$\frac{\partial^2}{\partial x^2} + \frac{\partial^2}{\partial y^2} + \frac{\partial^2}{\partial z^2},$$

which acts on a *eigenfunction* $\psi(x, y, z)$, and x, y, and z indicate the position of the electron with respect to the proton.

Operators differ from the functions that one studies in high school. Consider the function

$$f_1(x) = x^2.$$

If we select a value for x, the function is evaluated as a number. With $x = 2$, for example, $f_1(2) = 4$. An operator, on the other hand, "operates" on a function to obtain another (possibly different) function. Consider the operator

$$O = \frac{d}{dx}.$$

Letting O operate on $f_1(x)$ we find that

$$Of_1(x) = 2x,$$

which is a function that differs from $f_1(x)$.

Given an operator, it is interesting to find the functions that preserve their forms under the action of that operator. In a curious mixture of German and English, these special functions are called *eigenfunctions*. Thus for

$$f_2(x) = e^{ikx},$$

$$Of_2(x) = ike^{ikx} \propto f_2(x)$$

so e^{ikx} is an eigenfunction of the operator $O = d/dx$, and ik is the corresponding *eigenvalue*.

Whereas classical dynamic theory is based on functions of the spatial coordinates and time, quantum dynamics involves operators, and laboratory measurements reveal the eigenvalues of those operators. Thus in Equation (B.1) the square bracket

$$\left[-\frac{h^2}{8\pi^2 m_e} \nabla^2 + V(x, y, z) \right]$$

is an operator acting on the eigenfunction $\psi(x, y, z)$, and E is the corresponding eigenvalue.

In this operator m_e is the mass of the electron, and $V(x, y, z)$ is the electrostatic potential energy of the proton-electron system. Also h is called *Planck's constant*. This physical constant was introduced in 1901 by Max Planck in a paper on the problem of radiation from a black body that was a harbinger of modern quantum theory.

What does this all mean? Equation (B.1) looks rather like the classical equation for conservation of energy

$$\frac{1}{2} \left(p_x^2 + p_y^2 + p_z^2 \right) + V(x, y, z) = E, \qquad (B.3)$$

where p_x, p_y, and p_z are the components of momentum in the x, y, and z directions. Newtonian mechanics tells us that the x-directed momentum— p_x—is the mass times the x-component of velocity. Quantum mechanics, on the other hand, tells us to think of p_x as an operator. In other words

$$p_x = m_e v_x \rightarrow -i\frac{h}{2\pi} \frac{\partial}{\partial x}$$

and similarly for p_y and p_z, where "→" means "changes in the quantum theory to."

Equation (B.1) is a quantum mechanical statement of energy conservation, while Equation (B.3) is the classical version. The relationships between the two statements are these:

i) Equation (B.3) is an algebraic expression of energy conservation. Thus the sum of the kinetic and potential energies is a constant that can take any value.

ii) Equation (B.1) is an eigenvalue equation. It says that the action of the operator

$$\left[-\frac{h^2}{8\pi^2 m_e} \nabla^2 + V(x, y, z) \right]$$

on the function ψ is equal to ψ multiplied by the constant E.

iii) In Equation (B.1) E cannot take any real value. Only special functions—the eigenfunctions—satisfy Equation (B.1). For each of these eigenfunctions, $\psi_n(x, y, z)$, there will be a corresponding energy eigenvalue, E_n.

The frequencies in Equation (B.2) are given by Bohr's condition

$$f(n, m) = \frac{E_n - E_m}{h}. \tag{B.4}$$

Instead of skiing—or whatever—Schrödinger formulated Equation (B.1) and computed its energy eigenvalues to be (Schrödinger, 1926; Slater, 1960)

$$E_n = -\frac{R_{Th}}{n^2}, \tag{B.5}$$

where n is a positive integer. Expressed in terms of fundamental constants

$$R_{Th} = \frac{2\pi^2 m_e e^3 c^4}{h^2} \times 10^{-14} \tag{B.6}$$

electron-volts, where c is the velocity of light and e is the electronic charge. The spectral formula of Equation (B.2) arises naturally as the differences in energy eigenvalues calculated from Equation (B.5), and the theoretical value of the Rydberg energy can be compared with the experimentally measured value.

The most recent values for the physical constants are given in Table 3 (Cohen and Taylor, 1994). (In this table, the value given for the speed of light is exact because the meter is currently defined as 1/299,792,458 times the distance that light travels in one second.) The theoretical value for the Rydberg constant (R_{Th}) is compared with the corresponding experimentally determined value (R_{Exp}) in Table 4. The agreement is very good indeed.

Quantity	Value	Units	Error
Speed of light (c)	299 792 458	meters/second	exact
Electron mass (m_e)	$9.109\,389 \times 10^{-31}$	kilograms	$\pm\,.000\,005\,3$
Electron charge (e)	$1.602\,177\,335 \times 10^{-19}$	coulombs	$\pm\,.000\,000\,48$
Planck's constant (h)	$6.626\,075 \times 10^{-34}$	joule-seconds	$\pm\,.000\,004$

Table 3. *Most recent values for certain physical constants (from Cohen and Taylor, 1994).*

From Equation (B.1) it is seen that the time-dependent function

$$\Psi(x, y, z, t) = \sum_{n=0}^{\infty} c_n \psi_n(x, y, z) \exp\left(\frac{-2\pi i E_n t}{h}\right) , \qquad (B.7)$$

(where the c_n's are arbitrary complex constants), is a general solution of the *time-dependent Schrödinger equation*

$$\left[-\frac{h^2}{8\pi^2 m_e}\nabla^2 + V(x, y, z)\right]\Psi = i\frac{h}{2\pi}\frac{\partial}{\partial t}\Psi . \qquad (B.8)$$

From a comparison of Equations (B.5) and (B.8), classical energy becomes the operator

$$E \rightarrow +i\frac{h}{2\pi}\frac{\partial}{\partial t}$$

in the quantum picture. It is important to notice that the expression given for $\Psi(x, y, z, t)$ in Equation (B.7) is a sum or *wave packet* of time-dependent solutions of Equation (B.8). As was noted in Appendix A, a general solution can be constructed in this way because Equation (B.8) is *linear*.

How are we to understand the time-dependent wave packet $\Psi(x, y, z, t)$? Almost immediately after Schrödinger's first paper appeared, Max Born suggested—and the world of physics agreed—that $|\Psi|^2$ should be interpreted

Quantity	Value	Units	Error
R_{Th}	13.605 698	electron-volts	$\pm\,.000\,022$
R_{Exp}	13.605 698 1(40)	electron-volts	$\pm\,.000\,004$

Table 4. *A comparison of the theoretical and experimental values for the Rydberg energy.*

as a *probability density*. Thus

$$P(t) = \int_{x_1}^{x_2} \int_{y_1}^{y_2} \int_{z_1}^{z_2} |\Psi(x, y, z, t)|^2 dx\, dy\, dz \qquad (B.9)$$

is the time-dependent probability of finding the electron in the region of space defined by $x_1 \leq x \leq x_2$, $y_1 \leq y \leq y_2$, and $z_1 \leq z \leq z_2$.

Contrary to what we expect from our "common sense" or intuition—based, as it is, entirely on the classical picture—the wave packet of Equation (B.7) does not permit us to determine the values of momentum and position of a moving particle with arbitrary precision. If Δp_x is the uncertainty in the momentum and Δx is the uncertainty in the position, then Fourier transform theory tells us that Equation (B.7) requires

$$\Delta p_x \times \Delta x \geq h/2\pi, \qquad (B.10)$$

with identical relationships between Δp_y and Δy and between Δp_z and Δz. These restrictions on the accuracy to which one can determine momentum and position are called *Heisenberg uncertainty relations*. Another—that we shall return to in Appendix D—is

$$\Delta E \times \Delta t \geq h/2\pi, \qquad (B.11)$$

where ΔE is an uncertainty in energy and Δt is an uncertainty in the measurement of time.

Uncertainty in the position of an electron tends to grow with time as a result of the irreducible uncertainty in its momentum (speed) indicated in Equation (B.10). This uncertainty in the position can be reduced to its minimum value by an experimental measurement during which the wave function is said to "collapse" in a manner that is not described by the time-dependent Schrödinger equation or any other equation yet known. This is one of the mysteries of quantum mechanics.

The Born-Oppenheimer Approximation

onsider a molecule that is composed of N atoms. Assuming that the atoms, of mass M_k ($k = 1, 2, \ldots, N$), are standing still allows one to write the wave function in Equation (B.1) in Appendix B as the product

$$\psi(\mathbf{r}_j, \mathbf{R}_k) = u_{\mathbf{R}_k}(\mathbf{r}_j)w(\mathbf{R}_k). \tag{C.1}$$

Here $\mathbf{r}_j$ is a $3n$ dimensional vector that measures the positions of the electrons and $\mathbf{R}_k$ is a $3N$ dimensional vector for the positions of the atomic nuclei, where n is the number of electrons in the molecule and N is the number of atoms. (In H_2O, for example, $n = 10$ and $N = 3$.) When the atoms are allowed to move, Equation (C.1) remains approximately correct, with errors in calculating the energy eigenvalues of order

$$\text{Error} \sim \sqrt{\frac{m_e}{M_k}} \times 100\,\%.$$

This is called the *Born-Oppenheimer approximation* (Born and Oppenheimer, 1927).

Errors occur because the Born-Oppenheimer approximation assumes that the kinetic energies of the nuclei are zero, but in fact the ratio of kinetic energy of a nucleus to that of an electron is inversely proportional to the square roots of their masses. Thus in H_2O the ratio of the kinetic energy of a hydrogen nucleus (proton) to that of a valence electron is inversely proportional to the square root of the mass ratio of about

$$\frac{1}{\sqrt{1836}} = 1/43$$

or 2.3%, and for the oxygen nucleus this ratio is about

$$\frac{1}{\sqrt{16 \times 1836}} = 1/171$$

or 0.58%. These are irreducible errors that appear in exact solutions under the Born-Oppenheimer approximation, but many additional simplifying assumptions are necessary to find solutions of the Born-Oppenheimer equations so the actual errors are often much larger.

In Equation (C.1), $u_{R_k}(r_j)$ is a factor of the approximate wave function that describes the electron dynamics, and $w(R_k)$ is a factor describing the nuclear dynamics. These factors are calculated from the approximate Schrödinger equations:

$$\left[-\frac{h^2}{8\pi^2 m_e} \sum_{j=1}^{n} \nabla_j^2 + V(r_j, R_k) \right] u_{R_k}(r_j) \doteq U(R_k) u_{R_k}(r_j) \,, \qquad \text{(C.2)}$$

and

$$\left[-\frac{h^2}{8\pi^2 M_k} \sum_{k=1}^{N} \nabla_k^2 + U(R_k) \right] w(R_k) \doteq E w(R_k) \,. \qquad \text{(C.3)}$$

Equations (C.2) and (C.3) both have the general form of Equation (B.1), but the operator bracket

$$\left[-\frac{h^2}{8\pi^2 m_e} \sum_{j=1}^{n} \nabla_j^2 + V(r_j, R_k) \right]$$

in Equation (C.2) is an energy operator for the n electrons only. Notice, however, that the potential energy—$V(r_j, R_k)$—depends upon all the coordinates in the problem.

The operator bracket

$$\left[-\frac{h^2}{8\pi^2 M_k} \sum_{k=1}^{N} \nabla_k^2 + U(R_k) \right]$$

in Equation (C.3) is an energy operator for the N nuclei. The potential energy $U(R_k)$ for the nuclei is the eigenstate energy of the electronic system, and it depends only upon the nuclear coordinates. For a stable molecule (like water), $U(R_k)$ will have a minimum value when the electrons of the molecule are arranged in the lowest eigenstate (ground state) of Equation (C.2).

This minimum potential function

$$U_0(R_k)$$

is called the *Born-Oppenheimer potential energy surface*. Derivatives of this surface with respect to the components of R_k give the Born-Oppenheimer approximation to the forces acting on the atoms of a molecule. Thus if the components of $\mathbf{R}_k$ are written as

$$x_1, y_1, z_1, x_2, y_2, z_2, \ldots, x_N, y_N, z_N,$$

then

$$-\frac{\partial U_0}{\partial x_1}$$

is the force acting on the first atom in the x-direction.

For molecules with more than a few atoms, calculation of $U_0(\mathbf{R}_k)$ is a difficult numerical problem. Instead of calculating $U_0(\mathbf{R}_k)$, an approximating function is assumed to have the form (McCammon and Harvey, 1987):

$$\tilde{U}_0 \doteq \frac{1}{2} \sum_{ij} K_b (b - b_0)^2 + \frac{1}{2} \sum_{ij} K_\theta (\theta - \theta_0)^2$$

$$+ \frac{1}{2} \sum_{ij} K_\phi [1 + \cos(n\phi - \delta)] + \sum_{lm} \left[\frac{A}{r^{12}} - \frac{C}{r^6} + \frac{q_l q_m}{Dr} \right], \quad \text{(C.4)}$$

where the parameters are selected to fit a variety of experimental data. Typical sets of parameters in currently available computer codes have the following features:

i) The only interactions considered are between pairs of atoms.

ii) The first three summations are over pairs of atoms that share a covalent bond, indicated by the indices i and j. The first two are quadratic potentials leading to linear forces.

iii) The lengths of covalent bonds are indicated by $b = b_{ij}$.

iv) Bending angles of covalent bonds are indicated by $\theta = \theta_{ij}$.

v) Twisting angles of covalent bonds are indicated by $\phi = \phi_{ij}$.

vi) The fourth (last) summation is over pairs of atoms that do not share a covalent bond, indicated by the indices l and m. The scalar distance between atoms l and m is indicated by $r = r_{lm}$.

vii) The first two terms in the last summation describe a "six-twelve po-tential," which becomes large (i.e., repulsive) when the separation between the two molecules (r) is small and goes to zero for large r. With appropriate choices for the parameters A and C, these terms can represent either a van der Waals interaction or a hydrogen bond.

viii) The last term in the last summation accounts for electrostatic inter-actions between atoms. In this case the partial charges (q_l and q_m) must be estimated as well as the dielectric screening, D.

Local Modes

W hen is it really necessary to solve the quantum problem of Equation (C.3)? Under what conditions would a solution of the corresponding nonlinear classical problem be more convenient and accurate? Answers to these questions are suggested by an investigation of the C-H stretching vibrations in dichloromethane (Bernstein et al., 1990):

$$
\begin{array}{c}
\text{Cl} \\
\text{H}\longleftarrow\rightarrow\text{C}\leftrightarrow\text{H} \\
\text{Cl}
\end{array}
$$

A classical analysis shows that vibrational energy is concentrated on one (or the other) of the two C-H oscillators in a *local mode*. Thus one of the two C-H oscillators is vibrating with a large amplitude and the other with a small amplitude. This local mode oscillation appears because of the *nonlinear* nature of the classical problem.

Since it is *linear*, the quantum picture is different. Pure eigenstates do not share the classical local mode property because dichloromethane has a reflection symmetry between the two C-H bonds. The two lowest eigenstates are symmetric:

$$
\psi_s \doteq \frac{1}{\sqrt{2}}(|q\rangle|0\rangle + |0\rangle|q\rangle) \tag{D.1}
$$

and antisymmetric:

$$
\psi_a \doteq \frac{1}{\sqrt{2}}(|q\rangle|0\rangle - |0\rangle|q\rangle), \tag{D.2}
$$

where the antisymmetric eigenstate changes its sign if "right" and "left" are interchanged, and the symmetric eigenstate does not. The energy eigenvalues of these two eigenstates are separated in energy by

$$\Delta E = E_a - E_s \propto \frac{q}{(q-1)!} , \tag{D.3}$$

where q is the number of quanta in the combined oscillation. Now we can ask: How many quanta must there be in the oscillator before the dynamics of the quantum problem becomes indistinguishable from that of the classical problem?

Quantum theory approximates the classical behavior by mixing the symmetric and antisymmetric eigenstates in the wave packet

$$\Psi_1(t) = \frac{1}{\sqrt{2}} \psi_s\, e^{-2\pi i E_s t/h} + \frac{1}{\sqrt{2}} \psi_a\, e^{-2\pi i E_a t/h} , \tag{D.4}$$

which localizes energy on one bond (the left-hand bond, in this example) for a limited period of time. Equation (B.11) implies that it will remain localized for a (tunneling) time

$$\tau \equiv \frac{h}{\Delta E} \propto \frac{(q-1)!}{q} . \tag{D.5}$$

From its factorial—$(q-1)!$—dependence, τ grows rapidly as q increases. For $q > 5$ a classical calculation is in accord with experimental observations of the spectral linewidth. For $q > 21$, τ is greater than the age of the universe. If one is interested in dynamic behavior on a time scale that is shorter than the tunneling time, it is as accurate—and more convenient—to study a *nonlinear* classical system as a *linear* quantum system.

From another perspective, Equation (B.11) implies that whenever the product of the experimental resolution of energy multiplied the experimental resolution of time is large compared with Planck's constant (6.626×10^{-34} joule-seconds), it will be equally precise and much easier to use classical analysis.

Schrödinger's Cat

I s quantum indeterminacy biologically relevant? To answer this question let us take a look at Schrödinger's cat. This beast was dragged into the scientific literature to show that one should not use quantum theory beyond its range of validity and, in particular, in the realm of biology (Schrödinger, 1935), but some of Schrödinger's colleagues continue to disagree. Many believe that it is reasonable to write a wave function—ψ(cat)—for a cat, and, furthermore, that it makes sense to think about a wave function of the form

$$\psi_+ = \frac{1}{\sqrt{2}}[\psi(\text{live cat}) + \psi(\text{dead cat})]$$

(with energy E_+) to describe the experimental situation in which an experimenter is not sure whether a cat in a closed box is alive or dead. According to the standard rules of quantum mechanics, this wave function collapses into either ψ(live cat) or ψ(dead cat) at the moment a conscious observer opens the box and notes whether or not the cat is living. (Perhaps this conscious observer could be a vulture, who would eat the dead cat but leave the live one alone. Or a maggot.)

This "burlesque example"—as Schrödinger described it—can be discussed semiquantitatively by using Equation (D.5), which was derived to calculate the time interval over which the quantum behavior of localized vibrational energy of a simple molecular model is identical to the corresponding classical behavior. For a localized CH stretching vibration on a molecule of dichloromethane with more than twenty-one quanta, this time is longer than the age of the universe.

Applying a corresponding analysis to the poor cat we also consider a wave function of the form

$$\psi_- = \frac{1}{\sqrt{2}}[\psi(\text{live cat}) - \psi(\text{dead cat})] \, ,$$

with energy E_-. The difference between the two energies

$$\Delta E = E_+ - E_-$$

should be given by an expression something like Equation (D.3), where q is the number of elementary quanta necessary to describe a cat.

Assuming q to be the number of atoms in a cat ($\sim 10^{25}$) implies

$$|\Delta E| \propto \frac{1}{(10^{25})!} ,$$

a *very* small number.

To see just how small this number is, note that the time-dependent wave packet

$$\Psi_1(t) = \frac{1}{\sqrt{2}}\psi_+ e^{-2\pi iE_+ t/h} + \frac{1}{\sqrt{2}}\psi_- e^{-2\pi iE_- t/h} ,$$

corresponding to Equation (D.4) is equal to ψ(live cat) for a time

$$\tau \sim \frac{h}{|\Delta E|} .$$

Also, the time-dependent wave packet

$$\Psi_2(t) = \frac{1}{\sqrt{2}}\psi_+ e^{-2\pi iE_+ t/h} - \frac{1}{\sqrt{2}}\psi_- e^{-2\pi iE_- t/h}$$

is equal to ψ(dead cat) for the same time, where

$$\tau > 10^{10^{24}} \times \text{ the age of the universe .}$$

Long before this time has elapsed, the phase relationship that defines the quantum interference between ψ_+ and ψ_- will be lost through interactions with the environment. The cat is either alive or dead.

Since the cat cannot be considered "*gemischt oder verschmiert*" into a combination of living and dead states, it seemed unlikely to Schrödinger that quantum theory plays a role in the dynamics of life. One should not reject the possibility out of hand, because the maser, the laser, electrical superconductivity, and superfluids provide examples of quantum effects that are macroscopically observed, but those who suggest a quantum mechanical basis for life or for consciousness must shoulder the burden of proof.

In making a case for the relevance of quantum theory, theoretical arguments that show one *could* use quantum theory are not sufficient. One must show that it is *necessary* to use quantum theory because a classical picture is unable to describe experimental observations.

Linear and
Nonlinear Diffusion

I n a *space-clamped* measurement of the switching of a nerve
membrane (Cole, 1968), the membrane voltage, $v(t)$, is independent of the space coordinate x. The dynamics of this voltage
is then governed by the *ordinary differential equation*

$$c\frac{dv}{dt} + j_i = 0 , \tag{F.1}$$

where c is the membrane capacitance. Also, j_i is the ionic current per unit
length of the fiber, which is discussed in Appendix G. Since cv is the electrical
charge stored per unit length of the fiber, Equation (F.1) says that ionic charge
flowing across the membrane is balanced by the charge in the membrane
capacitance.

For a normally functioning nerve fiber, the membrane voltage is a function
of both time (t) and distance along the fiber (x). In this case, electric current
flows through the tube of the fiber in the x-direction, and $v(x, t)$ is governed
by the *nonlinear diffusion equation*

$$\frac{\partial^2 v}{\partial x^2} - rc\frac{\partial v}{\partial t} = rj_i , \tag{F.2}$$

where r is the electrical resistance per unit length of the fiber tube. Evidently
Equation (F.2) reduces to Equation (F.1) when v is independent of x.

Equation (F.2) is the fundamental equation of electrophysiology; it was
used by Hodgkin and Huxley to describe a nerve impulse in 1952. If the
ionic current j_i is zero, it reduces to the linear diffusion equation

$$\frac{\partial^2 v}{\partial x^2} - rc\frac{\partial v}{\partial t} = 0 ,$$

205

solutions of which can be understood from the following simple argument. Near the crest of localized (or "pulse like") initial conditions

$$\frac{\partial^2 v}{\partial x^2} < 0 ,$$

and this requires that

$$\frac{\partial v}{\partial t} < 0 .$$

Thus the initial pulse tends to flatten out with increasing time in a manner that is governed by the diffusion constant

$$D = \frac{1}{rc} \text{ cm}^2/\text{sec} .$$

If the ionic current is a nonlinear function of voltage, solutions of Equation (F.2) can be quite complicated. It is interesting, therefore, to look at special solutions that have the form of a traveling wave. Thus

$$v(x, t) = \tilde{v}(x - ut) , \qquad (F.3)$$

where the wave speed, u, is a free parameter in the calculation. Under this assumption, Equation (F.2) reduces to the second-order ordinary differential equation (Scott, 1977)

$$\frac{d^2\tilde{v}}{dx^2} + rcu\frac{d\tilde{v}}{dx} = rj_i , \qquad (F.4)$$

which can be solved by standard numerical methods for particular values of the parameter u. A traveling wave solution of the form given in Equation (F.3) corresponds to a nerve impulse, where u is its conduction velocity.

If P is the power (joules per second) dissipated by a traveling wave and E is the energy stored per unit length of the system, then the wave travels at velocity

$$u = \frac{P}{E} . \qquad (F.5)$$

It is at this speed that the wave "digests" energy (P) at the same rate it is "eaten" (uE).

It is interesting to note that in 1952—the same year that the work of Hodgkin and Huxley appeared—nonlinear diffusion was also suggested by Alan Turing as a basis for the development of form (morphogenesis) in biological systems. Once again we see a parallel relationship between the fundamental processes of life and those of mind.

The Hodgkin-Huxley Equations

I n the early 1950s Hodgkin and Huxley made careful measurements of the sodium and potassium components of membrane currents of the giant axon of the squid under space-clamped conditions, which implies that membrane voltage was not permitted to vary along the axon. From these data they constructed the following phenomenological expressions for the total ion current (Hodgkin and Huxley, 1952):

$$j_i = G_K n^4 (v - V_K) + G_{Na} m^3 h(v - V_{Na}) + G_L(v - V_L) \,, \qquad (G.1)$$

where

$$\frac{dn}{dt} = \alpha_n(1 - n) - \beta_n n \,,$$

$$\frac{dm}{dt} = \alpha_m(1 - m) - \beta_m m \,, \qquad (G.2)$$

and

$$\frac{dh}{dt} = \alpha_h(1 - h) - \beta_h h \,.$$

In these equations m is a *sodium turn-on* variable, h is a *sodium turn-off* variable and n is a *potassium turn-on* variable, and

$$\alpha_n = \frac{0.01(10 - v)}{\exp[(10 - v)/10] - 1} \,,$$

$$\beta_n = 0.125 \exp(-v/80) \,,$$

$$\alpha_m = \frac{0.1(25 - v)}{\exp[(25 - v)/10] - 1} \,,$$

$$\beta_m = 4 \exp(-v/18) \,, \qquad (G.3)$$

$$\alpha_h = 0.7 \exp(-v/20) \,,$$

Parameter	Value	Units
G_{Na}	36	millimhos/cm^2
G_K	120	millimhos/cm^2
G_L	0.3	millimhos/cm^2
V_K	−12	millivolts
V_{Na}	+115	millivolts
V_L	+10.6	millivolts

Table 5. Membrane permeability parameters for the giant axon of the squid. (From Cole, 1968)

and

$$\beta_b = \frac{1}{\exp[(30 - v)/10] + 1} ,$$

in units of milliseconds^{-1}. The membrane voltage v is measured in millivolts with respect to the resting potential, and an increase in the potential *inside* the axon is taken to be positive.

Values for the parameters in Equation (G.1) are given in Table 5. The physical scientist should be aware that the values of these parameters measured on particular axon membranes will exhibit normal physiological variation. Thus Table 5 describes a typical axon from which a particular axon may differ by 10% to 20%.

Equation (F.2) for nonlinear diffusion and Equations (G.1), (G.2), and (G.3) define the celebrated Hodgkin-Huxley (or H-H) system. Using these equations and assuming the solution had the form of a traveling wave, they were able to compute a numerical solution that agrees with experimental observations. In the wake of repeated testing over the past 43 years, the H-H equations have been established beyond reasonable doubt as fundamental equations of neurodynamics.

In Equation (G.1) the first (G_K) term accounts for potassium ion current and the second (G_{Na}) for sodium ion current, while the last term (G_L) accounts for all other ions under the heading of *leakage*. To understand how this system works, note that Equations (G.2) can also be written in the form:

$$\frac{dn}{dt} = -\frac{n - n_0(v)}{\tau_n(v)} ,$$

$$\frac{dm}{dt} = -\frac{m - m_0(v)}{\tau_m(v)} , \qquad \text{(G.4)}$$

and

$$\frac{dh}{dt} = -\frac{h - h_0(v)}{\tau_h(v)} ,$$

where, from Equations (G.3),

$$n_0(v) = \frac{\alpha_n}{\alpha_n + \beta_n},$$

$$\tau_n(v) = \frac{1}{\alpha_n + \beta_n},$$

$$m_0(v) = \frac{\alpha_m}{\alpha_m + \beta_m}, \qquad\qquad (G.5)$$

$$\tau_m(v) = \frac{1}{\alpha_m + \beta_m},$$

$$h_0(v) = \frac{\alpha_h}{\alpha_h + \beta_h},$$

and

$$\tau_h(v) = \frac{1}{\alpha_h + \beta_h}.$$

It turns out that sodium turn-on (mediated by m) is about an order of magnitude faster than potassium turn-on and sodium turn-off (mediated by n and h). Thus the switching of the membrane can be described as follows:

i) At the resting potential ($v = 0$), the sodium ion current is almost zero. This is because, from Equations (G.5), $m_0(0) = .053$ and $h_0(0) = .937$, so sodium ion permeability

$$G_{Na}m_0^3(0)h_0(0) = .00014G_{Na}.$$

The potassium ion current is also small at rest because $n_0(0) = .318$ so potassium ion permeability

$$G_K n_0^4(0) = .0102G_K.$$

ii) As the membrane voltage is increased from its resting value, sodium channels open ($m \to 1$) on a time scale of $\tau_m \sim 0.2$ to 0.4 milliseconds.

iii) This influx of sodium ions brings the membrane voltage to a level of $+115$ millivolts with respect to its resting value. Thus $(v - V_{Na}) = 0$ so the sodium ion current is also zero.

iv) At $v = +115$ millivolts, potassium ion permeability turns on ($n \to 1$) as sodium ion permeability turns off ($h \to 0$) on time scales of $\tau_n \sim 2$ to 5 milliseconds and $\tau_h \sim 1$ to 8 milliseconds, respectively.

v) Since $(v - V_K) = 127$ millivolts, potassium ions flow rapidly out of the axon. This efflux of potassium ions carries the membrane voltage back to its resting value on a time scale of a few milliseconds. It is important to remember that the sodium and potassium ion concentrations change little during this switching cycle.

Parameter	Value	Units
G_{Na}	0.104	micromhos
G_K	0.57	micromhos
G_L	0.025	micromhos
V_K	0	millivolts
V_{Na}	+122	millivolts
V_L	0	millivolts

Table 6. Membrane permeability parameters for an active node of a frog's sciatic nerve with an area of about 20 micron^2 (From Cole, 1968).

This behavior is not a special property of the squid axon. Table 6 presents corresponding parameters for the sciatic nerve of a frog. In this case the measurement was made on a small active node because each nerve fiber is largely covered with an insulating sheath (called myelin) to increase the conduction velocity of a pulse without increasing the nerve diameter.

Comparison of the data in Tables 5 and 6 shows that the properties of nerve from different phyla are quite similar.

Counting Neural Networks

Boolean arithmetic. The arithmetic of logic—which is used by digital computers—employs only two numbers: "0" and "1." The operations of addition "+" (or "OR") and multiplication "×" (or "AND") are defined as follows:

$$0 + 0 = 0$$
$$0 + 1 = 1$$
$$1 + 0 = 1$$
$$1 + 1 = 1$$

$$0 \times 0 = 0$$
$$0 \times 1 = 0$$
$$1 \times 0 = 0$$
$$1 \times 1 = 1$$

In early telephone exchanges, "0" represented an open relay contact, and "1" represented a closed contact. When Boolean arithmetic is used for logic, "0" corresponds to a false statement and "1" to a true one. In a modern digital computer, "0" indicates a low voltage level and "1" a high one.

Boolean functions are defined in the context of this arithmetic. The value of a function will be either "0" or "1," depending upon the values of A and B. Thus a particular function, $F(A, B)$, might be specified by the table:

A	B	F
0	0	1
0	1	0
1	0	0
1	1	1

and another, $G(A, B)$, by the table

A	B	G
0	0	0
0	1	1
1	0	0
1	1	1

General Boolean nets. The most general net with loops of logical feedback is a system of N neural elements, each of which can compute an arbitrary Boolean function of the output states (1 or 0) of all N elements. There are 2^N possible states of an N-element network, and each input code requires that the output be either a "0" or a "1." For two elements with two inputs each, for example, the four input codes are $(0, 0), (0, 1), (1, 0)$, and $(1, 1)$. A particular Boolean function is specified when the output is determined for all four inputs. How many such functions can be constructed? To answer this question, consider that the first input code $(0, 0)$ must require the output to be either a "0" or a "1." Those are two possibilities. For each of these possibilities there are two possibilities for the second input code, $(0, 1)$, giving $2 \times 2 = 4$ arrangements for the first two input codes. Carrying on, there are $2 \times 2 \times 2 = 8$ arrangements for the first three input codes and $2 \times 2 \times 2 \times 2 = 16$ arrangements for all four input codes. Thus there are

$$2^{2^N}$$

Boolean functions that can be constructed for the 2^N input codes of the general case.

Next we consider how many systems (or logical machines) can be constructed from N Boolean elements. The first element can be selected as one of 2^{2^N} functions, and for each such selection the second can also be chosen from 2^{2^N} possible functions, and so on. Thus the total number of such systems is

$$\#(N) = \left(2^{2^N}\right)^N = 2^{N2^N},$$

a number that grows rapidly with N. For $N > 6$, $\#(N) > 10^{110}$, so there is an immense number of general Boolean systems for $N > 6$.

General threshold nets. From the discussion of Chapter 5, we should not expect a biological neuron to compute *all* the Boolean functions of its inputs. For McCulloch-Pitts neurons it has been shown that the number of Boolean functions that can be computed by each neural element is (Yajima et al., 1968)

$$\sim 2^{kN^2},$$

where k lies between one-half and one, so the total number of systems is

$$\#(N) \geq 2^{N^3/2}.$$

This implies an immense number of systems for $N > 9$.

A conservative estimate for the number of human brains. Considering the large number of neurons in the brain, it is not realistic to suppose that every neuron receives inputs from all others. If the number of inputs to each neuron is n, the total number of systems is

$$\#(N) \geq 2^{n^2 N/2}, \qquad\qquad (\text{H.1})$$

which implies an immense number of systems for

$$n^2 N > 731.$$

Since $N = 10^{10}$ is a conservative estimate for the number of neurons in the human neocortex and $n = 10^4$ is a reasonable estimate for the average number of synaptic inputs to each neuron (Stevens, 1993),

$$n^2 N \sim 10^{18},$$

which is much greater than 731. From Equation (H.1) the corresponding number of mental systems is

$$\sim 10^{10^{17}},$$

which is greater than $\Im^{10^{16}}$, where $\Im = 10^{110}$ as defined in Chapter 2. This is a *very* large number: the immense number multiplied by itself ten thousand trillion times. The numbers used by astronomers are minute in comparison, and it should be emphasized that this estimate was made under the most conservative assumptions about the dynamics of a nerve cell.

An Incomplete Physical Theory

s a simple example of an inadequate theory, Wigner introduces Maxwell's equations for electrodynamics in free space (Wigner, 1969):

$$\frac{\partial E}{\partial t} = c \operatorname{curl} H,$$

$$\frac{\partial H}{\partial t} = -c \operatorname{curl} E,$$

$$\operatorname{div} E = 0, \text{ and}$$

$$\operatorname{div} H = 0,$$

where E and H are the electric and magnetic field intensities respectively, and c is the velocity of light. These equations can be written as a second-order equation for the electric field alone. Thus

$$\frac{\partial^2 E}{\partial x^2} + \frac{\partial^2 E}{\partial y^2} + \frac{\partial^2 E}{\partial z^2} = \frac{1}{c^2}\frac{\partial^2 E}{\partial t^2}, \tag{I.1}$$

where $\operatorname{div} E = 0$.

If we knew Equation (I.1) but were ignorant of the existence of the magnetic field intensity H, we could correctly calculate the evolution in spacetime of the electric field intensity E but would have difficulty in computing the dynamics of a charged particle, which depends on the magnetic field. To effect this we would need the additional equation

$$H(r) = \operatorname{curl} \int \frac{\partial E(r')}{\partial t} \frac{d^3 r'}{4\pi c |r - r'|}, \tag{I.2}$$

without which the theory of electromagnetism is incomplete.

215

References

R Ablowitz. The theory of emergence. *Philos. Sci.*, 6:1–16, 1939.

L Bernstein, J C Eilbeck, and A C Scott. The quantum theory of local modes in a coupled system of nonlinear oscillators. *Nonlinearity*, 3:293–323,1990.

M Born and J R Oppenheimer. Zur Quantentheorie der Molekeln. *Annal. Physik.*, 84:457–484, 1927.

E R Cohen and B N Taylor. The fundamental physical constants. *Physics Today*, Part 2, August 1994, 9–13.

K S Cole. *Membranes, ions and impulses*. University of California Press, Berkeley, 1968.

A L Hodgkin and A F Huxley. A quantitative description of membrane current and its application to conduction and excitation in nerve. *J. Physiol.*, 117:500–544, 1952.

J A McCammon and S C Harvey. *Dynamics of proteins and nucleic acids*. Cambridge University Press, Cambridge, 1987.

E Schrödinger. Quantisierung als Eigenwertproblem. *Annal. Physik.*, 79:361–376, 1926.

E Schrödinger. Die gegenwärtige Situation der Quantenmechanik. *Naturwissenschaften*, 23:807–812, 823–828 and 844–849, 1935.

A C Scott. *Neurophysics*. Wiley, New York, 1977.

J C Slater. *Quantum theory of atomic structure*. McGraw-Hill, New York, 1960.

C F Stevens. Reworking an old brain. *Curr. Biol.*, 3:551–553, 1993.

A Turing. The chemical basis of morphogenesis. *Phil. Trans. Roy. Soc. (London)*, B237:37–72, 1952.

E P Wigner. Are we machines? *Proc. Amer. Philos. Soc.*, 113:95–101, 1969.

S Yajima, T Ibaraki, and I Kawano. On autonomous logic nets of threshold computers. *Trans. IEEE Computers*, 17:385–391, 1968.

Index